Quality for Intelligent People

by Bogdan Cojocaru

Contents

Short preface   1

How should you understand Quality?   2

The unseen and often forgotten attributes of Quality   9

Know your user   23

The natural enemies of quality   32

State of the art   49

Quality of life   58

When did you last manipulate your customer?   66

Tester's amusement park   77

The rotten apples   91

Short preface

After close to almost a decade of experience in Quality Assurance and Quality Control, I felt the need to summarize some of the essential ideas about quality that I learned through my own experience. I didn't want to turn this into a manual for testers, but instead make it an easy reading for all those interested in the subject. From my perspective it should be read mostly by managers, marketers and anyone who wants to have a global view on the essence of Quality.

I consider it a holistic approach to quality, reuniting ideas coming from multiple domains – from psychology, sales, software development, marketing, advertising and even philosophy. In the end all knowledge is interconnected.

As an official disclaimer: this book presents my personal views based on my own experience and does not represent the perspective on quality, or any other topic discussed, of any of the companies or clients that I worked for.

Since I never was a fan of prefaces and I usually skip them when I read a book, I will follow the KISS principle (**K**eep **I**t **S**hort and **S**imple) and stop here, inviting you to start reading the first chapter.

How should you understand Quality?

Theory offers a number of definitions for quality that are of little practical use when it comes to actually understanding what quality is about. I would offer you four alternatives which can be applied and remembered in many contexts:

Quality means fit for purpose

If you would place winter tires on your car and drive it around on sunny dry roads or in autumn temperature, you will just notice a heavier consumption and a slower car. To be able to appreciate their quality you need to test them on heavy snow or icy roads, on a low temperature, or at least on a cold and rainy weather. Of course this is an example everyone gets, more or less, but what happens when you configure a computer for let's say audio processing and then you try to run the latest video card-challenging, computer games on it? It will fail like a surfer at the winter Olympics.

There is only one answer to the examples above: it wasn't designed for that. Question is: does your user know it? To optimize sales, most companies try to present their equipment as performing excellently on any kind of "weather". There is no such thing! Being good at everything means not performing in a specific direction.

Products made for all-season, all-weather, all-purpose, all-anything, although created as a response to the need of having something of general use at an affordable price, end up being the low standard of good quality.

You will not encounter a hi-tech device that's made to endure the harshest conditions: water resistant up to 50 meters, dust proof resistant, rock smashing resistant and still being able to take excellent audio/video capture, plus an excellent data signal. It's not because it wouldn't be possible, but because it's not efficient: you don't have a market big enough to justify the cost of producing such equipment at an affordable price, to justify the allocation of resources. Not to mention some features will automatically impact the design.

It would be an investment in features that most consumers, as the majority of city inhabitants for instance, don't actually need. Besides, companies don't want you to make your next buy in twenty years from now, just because the product they created is extremely resistant. Moral depreciation is not a reliable factor to measure the number of returning costumers, either.

I remember an example from the beginning of my career when I was working in customer support for a major electronics producer. A client called in to complain that his camera, which he had dropped a few days before in the water (staying there for a couple of days), once dried, was malfunctioning. Believe it or not, the main problem was that the flash worked no more. Needless to say this was beyond any reasonable warranty. In fact that camera exceeded the purpose it was designed for by still being able to take pictures after being submerged in water for three days.

Designing a product fit for purpose will not be enough, because:

Quality is in the eye of the user

This is the second rule that helps you define quality. No matter how much effort you put in to justify that 2% advance/ improvement compared to previous version, if the end-user can't see a change, you've done nothing. If you expect the end-user to pay more for something he can't see or feel in any way, you're not going to get his money.

Remember those updates you get on your phone or on your tablet, which say: although it looks exactly the same, it's much more stable now. From the psychological point of view, if the user can't see a difference, there is no difference (and the customer is always right! ... in 7 cases out of 10). If that annoying error he was running into every day, because you've done some lousy coding is not removed, then it's not more stable. Naturally, for computer literates, "stable" has a different meaning (nothing to do with horses either) and regular updates make sense, but the average consumer - which is the end-user for most mass products - is often a tech rookie. Companies that have realized this put a small effort into changing at least the icon and surprisingly, trivial as it may seem, it makes a difference: something visible has changed. Rounder corners, different texture, a change in the color theme, new sounds – these are the kind of small changes that have no contribution to upgrading the product, yet they matter. Do not mistake this for a facelift which is a major redesign and influences sales.

I will share with you another story that is 100% authentic, but due to non-disclosure agreements, I will not share any names or specific data. A new version of camera firmware was released integrating a feature of face detection. Like most things in this world, it wasn't perfect. It didn't reach

100% efficiency in detecting faces and we're still at least a decade away from reaching that point. Prior to demo, during regular testing phases, the product was tested over a huge database, testing different race, age and sex distribution to name just a few criteria. At the moment it was considered to be the most complex database in the industry.

Among the audience of this demo, there was a CEO, who picked the camera at the end of the meeting and later, same day, took a photo of his little daughter. For that particular face of his lovely little one, the detection failed: it didn't work. The CEO was very upset about this, though it was honestly admitted from the very beginning that efficiency is not 100%. The negotiations didn't rely on one man's subjective opinion, but it sure made getting the contract a lot more difficult. For that particular end-user, the product was simply not worth it, as it wasn't covering what was most dear to him: his daughter.

Third definition is actually adapted from a quote attributed to Henry Ford:

Quality is what you do when no one supervises you

Or in more common language it means not sweeping the dust under the carpet, if no one is watching. Quoting Mr. Ford himself: *'Quality means doing it right when no one is looking'*.

This 'guideline' often links to the phrase: *'oh what's the point, no one will notice it anyway'*. Instead of assuming no one will notice, ask yourself the opposite: what if someone will notice it? Will you feel embarrassed? Will there be consequences of any kind? Will your reputation be irremediably affected? If not, then go ahead: sweep the dust under the carpet. If however the answer is yes to any of these questions or you care about your reputation, think twice before bringing your contribution to the huge pile of junk that already exists in the world.

I remember buying a desktop computer for someone many years ago (yes, it was that long time ago, when desktops were still a popular choice). Despite a passion for putting the hardware together and configuring it myself, I didn't have the time and the budget didn't allow for a spectacular configuration, so there wasn't that much fun in working on it. But to bring it a personal touch, I wanted to add a few extras, change a few jumpers on the motherboard and so on. Now remember that this was a standard configuration, from which I started. They were putting together dozens like that one, so normally you'd say no one has time for details. Furthermore, being an average configuration the target audience wasn't that demanding – you would not expect the buyers to break it open and see what's inside.

Well, when I opened the desktop case, those guys at the computer shop won my respect: all the components were perfectly put together, no loose wires hanging around, they were all tied up together, put aside, not to block the ventilation and to allow access to components. There was nothing left for me to do at the hardware level. They did the job right, even though no one was watching. Of course, maybe I was lucky, maybe there was just one guy in their assembly department working so neat, but if so, that is the kind of guy you ought to propose for a retention bonus, because it made me assume this is how they usually work. I guess we are more familiar with the opposite situation when one employee treating you badly leads to labeling the whole business with the sentence: *'those guys are idiots'*.

People easily get these day-to-day examples, but it's harder to take it out of the context and apply it when working with documents or processes: writing junk in strategies because nobody has the time to read them anyway, not commenting the code, because if someone needs to understand it, *"we'll just do a quick handover"*, not updating the history of a document, because for now 'track changes' feature is activated, not updating the user manual, because people call customer support anyway. Except for hardware user manuals, where most products – even upgraded versions – are treated as individual new products, all other manuals I have read (from software to corporate instructions) were so outdated (I'm talking up to 5-7 years here) that they were of little practical use.

Remember that it's in the human nature to remember those who disappointed us, rather than those who helped us.

Quality is predictable.

This is one definition from the manufacturer / producer's perspective.

Quality does not occur by accident and should not be a surprise. If you hold the recipe for it, you know and expect the outcome. If it's predictable it means you know what took you there and you can do it again: success is also repeatable. It means you know your user (we have a separate chapter for that), the market, your business and you're not making any friends with '*the natural enemies of quality*' (which we'll discuss also in a separate chapter).

Last, but not least, I want to include here a definition that has a seed of truth in it, but did not make it to my top four, because it's easily disputable:

Quality means superiority

Up to a point this is correct. If we compare a product to a dozen other products and it performs better, we say it's superior or "good quality". Fair enough. But what if that reference system is just a dozen of mediocre products? Better than mediocre may still mean bad. I dismiss this description of quality, because I don't think quality should be interpreted by comparison, but should be analyzed in its individuality and for that we can apply the four principles listed earlier. For the sake of entertainment, I want to remember here the electronics producers who at some point were manufacturing Walkmans and the moment they introduced cd-players, they started to thrash around (or at least discredit) in ads the products they were making themselves the other day. Until they invented the mp3 player and then they despised cd-players.

Given the speed at which technology evolves, calling something superior is simply amusing. Catapults used to be superior military equipment. Incantations were at some point a superior form of medicine. The legendary Nokia 3310 was at some point superior. Superiority is a temporary attribute. But embedding in a product *the unseen attributes of quality*, can bring it to a *state of the art* level.

The unseen and often forgotten attributes of Quality

Most good books about Quality Assurance will tell you about non-functional requirements like robustness, scalability, maintainability, accessibility, mean time to recovery, security, privacy, safety, etc. I am really glad they do, because this book is not a testing manual and I truly hope everyone interested in Quality already realizes that reliability, portability, security, etc are important. The above features will give you a good, well appreciated product, will allow you to enter the 'first league', but will not make it 'THE product'. This book wants to make you aware of those harder/impossible to quantify attributes that can take your product from the first league, to the dream team. You mustn't ignore the first traits of a good product that I mentioned earlier, but these are the add-ons which any exceptional product should have. So let's see what else we can add...

Beauty / Esthetics

Yes, exactly. Beauty and esthetics are terms often associated with cosmetics or make-up. They pop-up in discussions about design (industrial design or just fashion), but you don't hear them mentioned too often when it comes to software or other products. Let me ask you something: have you seen apps that are plain ugly? So ugly that they have no excuse for their own existence?

People working in finance or large corporations have to suit up every day and spend at least eight hours each working day (sometimes weekends as well) in an environment that's purely depressive. The inventor of cubicles should be stapled onto a wing panel (wall of a cubicle). Having to work most of those eight hours with a software that's all blue and gray, all straight lines and table-looking, enemy of every round shape, all this adds up to whatever makes people hate their job. Blue, gray, straight lines, Arial font, black background or SAP-like skin are considered to make things look serious. Accounting is serious enough already, but apps made for book keeping are dead-serious. The only curve line you see is probably part of a logo. Think about the end consumer for a second and give that person a helping hand.

I've seen on the other hand software applications ending up like this, due to an excess of compliancy to accessibility standards. That means making the software accessible for physically challenged persons. As much as I support this initiative and enforce it (particularly when it comes to a lifesaving product – for example a device for emergency calls), a 100% compliant software product would be all black and white, with big buttons and embedded audios. Better option would be to create a special edition or optimize the existing version, which would take some extra months of work, that in turn translate to a lot of extra cost – which let's face it, nobody is willing to actually spend. Caring for the others or covering words with facts is much easier, if it's free.

For those of you who think that these regulations or guidelines are easy to follow, to give just one single example, I should mention that they define even the contrast rate that should exist between two elements, including the case where the page would be displayed in black and white. I want to underline once again, that I perfectly understand the necessity of these features and I support any initiative to make products compliant with these regulations, but most companies are doing it just because they have to, without any consideration for the end-users.

So what about the colors? "Colors are for women" is a common stereotype, yet every stereotype has a seed of truth in it. Whoever went shopping for a mobile phone with his wife or girlfriend, probably heard the strongest argument: 'I prefer that one, cause it's pink' or 'I'll take this one, cause it's got a butterfly design on the back'. Real life examples folks, I'm not sexist. The producers of these products got the point: they know their target audience. Marketers should not be led by stereotypes, but they should try to understand them and well...exploit them.

In the end, Esthetic sells. Harder to accept when it's called 'beauty', much easier to accept when it's called 'design'. Take for example an ordinary brown carton box. Let's say you sell it for a dollar. Put a leaf drawing on the sides and it becomes 'brown carton box with leaf design' – price: 3 dollars. Make it white glossy carton, with green leafs, chromed edges, market it as 'Nature's hideout' and you can sell it as a deposit box for 15$. In the end, it's still a box.

This area is maybe easier to exemplify by bringing in negative examples of companies that failed. Here we find in the auto industry the former/future car producer SAAB with its Automobile division, which in 2011 declared bankruptcy. Of course, SAAB group continues to exist, even the Automobile division went on, under a different name, with different shareholders, but this is of little interest for us. This group was accustomed to produce airplanes and plane engines, before cars, so naturally they thought of giving it a shot in the automotive industry. Without discussing the price, which is often prohibitive for many people, I find a company that builds fighter jets and military defense systems to be competent enough to make a car that runs on a highway. But if you make a search on the Internet for 'world's ugliest cars' – SAAB is present in nearly every top.

Why did this aspect matter so much? Ugly? Because people try to make their car a reflection of their personality, by picking a particular design, or at least a particular color or different accessories. No one wants the car to say about him, or especially her: *'look, I am damn ugly'* according to most people's standards. On the other hand one of my favorite cars in terms of design – the Chrysler PT Cruiser is considered by many the ugliest thing on the roads. Of course this will not stop me from buying it, in case I ever decide to do so, but it could prevent a lot of other people from choosing it.

Ok, so we got the connection to sales and marketing, but how does it affect the actual quality? A piece of furniture will not break easier just because it has no ornaments, or sense of design. Then, what's the catch? As I've said in the first chapter: quality is in the eye of the user. Maybe some adjustable metal shelves holding up to 100kg each are fit for the purpose of holding tools in a garage, but making them look less like Lego pieces and more like mobile or modular furniture, would sure win you a few extra buyers and more than a few extra bucks.

User-friendliness / usability

Ask yourself an honest question (and hopefully you are in a high enough position in your company when you ask this): is your product the one you/users would prefer over your competitor's?

In my opinion, one of the biggest wins of Google Chrome's browser was the fact that you could enter a search in the address bar. This was responding to two nuisances. First one: whenever you enter an incorrect site address, older browsers will just give you the answer 'This page does not exist' – very helpful indeed. So in that case, you can try again if you remember the correct address or explicitly open a browser, go to a search engine, then search for the key words. Chrome was giving you options: either you type it as a web address, we'll tell you it doesn't exist, but give you one button to turn it into an instant search OR just type the name of the website, without www in front and without the domain, we'll do a search and if your guess was correct, that website will be the first result in the hitlist.

Second nuisance coming from the other browsers, was the fact that you had to open explicitly a browser page to do a web search. Of course, having the most popular search engine behind Chrome made a difference, but all these facilities could have been easily integrated even in IE with a bit of extra coding to make it a more pleasant experience. Next things that conquered me were the fact that I could open multiple tabs in the same window and close with only two clicks, all pages but one, or just close the tabs to the right.

Let's see another example.

One of the user-friendly features coming in Audi cars is the fact that they lower the windshield wipers, when not in use, so that they are completely off-sight. Having them left there as in most cars, laying at the bottom of the windshield wouldn't have affected one's visibility or driving, because nobody watches the road through that area. But they thought: *"hey, maybe they annoy you, just hanging there"*. Maybe you just want a clear sight of the beautiful landscape in front of you. *"We'll just take them out of your sight completely, until you need them."* Refinement is in the details. These are the things you remember when you don't have the product anymore or when you have to use a replacement, thus making you appreciate it.

Responsiveness to change, often placed under the umbrella of maintainability is in my opinion part of what make a product user friendly. The more expensive the product, the more adaptable it has to be. Hardware products must have the latest ports available and be able to deliver new features via firmware update. You can't expect a customer to pay a high price or even a medium price for a product that's completely history in one year. Official excuses like 'production has been discontinued' or 'no longer supported' will not do. That's just garbage and everyone knows it. As long as it is on the shelf, people expect to be able to use the product they've just bought for at least a few years and be glad they have it. The drive to replace should be the moral worn out – the fact that completely new technology has emerged or something new was invented that people want to have.

To give you an example: same manufacturer – let's call it Eastern Analog just to give it some random name - in the same year had on the shelves a media box and a portable hard drive. The media box, was not able to play most HD videos and also didn't support the most popular audio codecs for HD video files. It was a pure firmware problem which could have been easily solved by regular updates, which got discontinued. Next version of the media box came in with regular updates and a bunch of plenty other features as it had wi-fi connection and network support. The sole presence of wi-fi and LAN ports would have been enough to attract customers, but they would rather drive crazy their old customers, by forcing them to buy the new model. Making people angry was not a smart move in an era where most TVs support memory sticks and laptops come with an HDMI port.

As I was saying, the same manufacturer had at the same time a portable hard drive on sale. This was happening when portable SSDs were still expensive. They were offering a big storage capacity, USB connection, the only drawback being the size of this 'portable' drive which resembled in size a novel by Tolstoi. If you cracked open such a drive, you would find out that inside it was a regular internal SATA hard disk (like the ones you would find in a desktop computer), adapted to support an independent power source. Not quite a big surprise from the technical point of view.

However, this thing remained fully functional and usable, for years, driving users to get the new model just by its size. Inside the plastic case, there was a lot of space outside the essential materials, just enough to make enquiring minds wonder whether there was a heat distribution problem, a pure design choice or a marketing approach that had its moments of glory for a short while: get the novelty out in an ugly as **** case and people will buy out just because it's new. Then make version 2.0 with the exact same content, but in a new case, and the rest will buy it because it looks better (and can throw it into their faces to those who already had version 1.0). What actually drove the decision for the initial design, we'll never know. Or not share it.

Some years ago, a chain of gas stations implemented a card system for their faithful clients, based on which clients could fill their tanks any time without paying on the spot and instead pay at a particular date, each month. A sort of 'gas on credit'. Card holders had to present this card whenever adding fuel (by swiping it through a reader at the gas pump), but if one chose the 'regular client' option, wanting payment by cash or credit card, this option seemed impossible. The system was not releasing the electronic lock on the pump mechanism, plus the order was not being transferred to the cashier (since there was nothing to pay for) – you will see why. The customers that didn't have this card (which was optional anyway) – most of them not regular users of these stations – had to press a button before anything else.

What the owners didn't take into consideration, was the natural sequence of actions (or possible state transitions), because if you were lifting the hose before pressing the button or trying to put the hose back and press the button again, you could not exit the 'card holder client' flow. This caused a lot of frustration among these occasional users, which decided to stop using their services. In fact they were impacting also the use case where a card holder might've forgotten his card at home and wanted to pay cash.

In the early days of Agile development, the Agile manifesto was stating: "(we value) working software over comprehensive documentation". Working software is something opposite to the system described earlier. A small note for younger people developing software: when the Agile manifesto came out, they were fighting mostly with the Waterfall software development model, as the V and W models although known and used for some years, were still stretching their roots. The wisdom behind this statement is that if the software is indeed user friendly, nobody needs to read the manual.

As customer support people know, even if people DO need to read the manual, usually nobody does it, hence the RTFM rule in customer support. Also, I never understood why some companies are treating the user manuals as if they were the Dead Sea Scrolls, making it impossible to find them, particularly for discontinued models. Are they so embarrassed about what they created at some point? Do they actually believe someone is going to steal their obsolete technology by reading an old manual? Couldn't they make a 'Museum' section where you can find the pdf-s you need?

Doesn't matter. As I said: for someone not to need the manual, the application or product must be user-friendly. If you want to know how user-friendly your application is, take this test: ask yourself (and answer honestly) one simple question: *"How would you describe it to a blind person?"* Easy to understand is easy to describe and a picture is worth a thousand words, but if your listener can't see that picture, would you need a thousand words to describe it? If you have the time to extend this exercise, ask someone to draw what you describe and go through the features of your product: tell the story of how you execute some particular action.

Needs come first, profit comes second

A good product answers a need or a problem. That's why we're not making screwdrivers or hammers with embedded Bluetooth or mp3 players. Adding features that are of absolutely no relevance for the end product itself, is not adding any value. Profit follows naturally any good product, with a healthy lifecycle and distribution system. Trying to create artificial profits by adding features nobody needs, is not going to work. Running shoes with LEDs were 'cool' for a 5-year old, tops and so were perfumed notebooks.

Pre and post-delivery services

The marketing department is well aware of the importance of these two factors, but how can they impact the quality of the actual product? Well, as I pointed out in the first chapter, quality is in the eye of the user and the end customer does not always distinguish between delivery services, manufacturing services, servicing and warranty (sometimes the customer can't even know all these aspects).

Canon for example considers itself a manufacturer. They don't sell directly to private persons, only to resellers. Therefore, when it comes to warranty, the end customer cannot address directly to Canon, but to the reseller. Forget for a second the name 'Canon' because this is just an example. There could be any other name instead and any other number of services outsourced or legally split between different entities. Let me ask you a simple question: if the end user has a problem with the product (perfectly covered by warranty and explainable when you are producing millions of units), do you think that person would like to hear that you are just the manufacturer and he/she should complain to whoever sold him the product?

A more common example would be pizza delivery. You are ordering a specific pizza from a specific place, because you like it. BUT part of the service is the actual order-to-delivery time and sometimes the post delivery – which ranges from granting fidelity bonuses to dealing with the situation when they deliver the wrong order. So, you're making the best pizza in the world, but delivery is slow (and hungry people are not getting friendlier), the delivery guy doesn't have the right change and when the customer calls back because he got the wrong order, you say he can't receive his money back. Hopefully, it's highly unlikely to deal with all these three situations at once, but do you think the customer will care whose fault it is? No, he will simply not order again from you.

It's perfectly reasonable to outsource or contract different services which you can't or don't want to cover. This doesn't mean you've washed your hands away altogether. Pre and post delivery services, along with the product itself create the overall client experience. You can't have a truly excellent product, with zero quality services, because the chain cannot be stronger than its weakest link (we're going to discuss in a later chapter a bit about the Theory of Constraints).

This part can be covered through internal incentives (to make sure your sales personnel or customer support has a positive reason to put a smile on) and by facilitating user complaints regarding outsourced or external services, but also having SLA-s in the contract with these third parties.

Pure design

I kept the design as a separate trait, being different than esthetics which is entirely subjective, as perceived by a group of users. Pure design has to do with ergonomics and texture, unity, balance, contrast. I am in no position to give lectures about design, but it has always fascinated me.

So let's take two examples that will help us in our pursuit for quality.

When the iPhone was preparing to take over the market, during its design phase, a particular attention was given to the round corners and to the texture of the case. We all had the sensation at some point in our lives, while using a phone or holding an object that the texture felt like cheap plastic. This is something you want to avoid when making a product. Curved mobile phones are not something new, but touchscreen elements have put their constraints on ergonomics for a number of years, so that only a good number of years after the release of the first touchscreen phones, the market was able to see curved phones once again.

Ergonomics are highly visible on keyboards and peripherals. Mouse controllers have adapted a lot to find the right curve to match the human hand, the right type of sensor that offers accuracy but also shortens the movement. Keyboards have also transformed to offer a comfortable hand position, have lowered and curved the keys to make it feel more natural. Of course we could live without these elements, functionally speaking, but since the examples I've picked are items that you use about eight hours per day in an office environment, that little touch of design that makes it more pleasant and healthier (!), matters a lot.

This is the useful part of the design, where through ergonomics your body can literally feel the benefits of a good design. Think of a desk chair. I once worked in an office, where the standard chairs were such an object of torment that I could literally feel a metal bar punching me in the back every time I was trying to lay back. They were impaling instruments. These chairs were of a terribly poor quality, quite embarrassing I would add, as they were breaking one by one in less than 3 months of use so it didn't take long until I decided to bring my own comfortable desk chair (soon followed by many others).

In design, there is also the visual part, **the eye candy**, which makes a user appreciate a product. A chrome coating can make things look classy or expensive, a carbon fiber look may suggest endurance, resistance, naked mechanisms are meant to show the technology within. At some point, I wanted to buy a wrist watch with LED display, which had an unusual algorithm for displaying the time and date (using a combination of dots with a particular distribution), twisting your brain a bit to read it, but I appreciated the concept. Of course, as they moved away from simplicity and the notion of jewelry associated with watches, they've cut substantially on their market. Yet what made them lose another bunch of customers were the childish strong LED lights they used, looking like Christmas tree lights, purely embarrassing for an adult. A neat option could've been a fade effect or dim lights, but no, they had to throw the whole concept down the drain at the end.

Uniqueness

"*In order to be irreplaceable one must always be different*" said once Coco Chanel – a world famous designer. I see it valid for products/services also. If you're not bringing anything different in your product, by comparison to competition, why wouldn't I, the customer, buy another company's product?

All you need to do is find design / implementation alternatives for each feature, then try to combine them in a unique fashion. No one is asking you to reinvent the wheel – but it would be great if you could do so. There are always classical safe options available, leaving you with the simple task of selecting the components. Once you found a unique pattern that defines what you do, you are starting to shape your brand.

If you want to come out with an original design, you must know your target audience. This brings us to our next chapter.

Know your user

Brands like Lamborghini or Ferrari sell first of all an image, besides a super car model. You need a perfect road to drive it around, as even a twig laying on the asphalt could scratch your front bumper. An average person would also need a second salary just to buy gas.

So let's say you're making speed cars. Two seats, three doors and the third one usually doesn't expose the trunk to throw in groceries, but the subwoofer and audio system to blow up party people's minds. Obviously this is not a family car, so you can't bet on or promote space or safety (although safety is highly important on a car that's supposed to reach 100kmh in a few seconds). So, can you perceive quality through the eyes of these consumers? You can't decorate the interior of a sports car like a Bentley using wooden finishing and leather (I'm mentioning leather just because in the 21st century, many people still consider it stylish to use the skin of a dead animal).

These consumers – your sports car buyers- will appreciate breaks, ease of changing gears, acceleration, suspensions, an excellent audio system and so on, first because they usually know a thing or two about mechanics and / or electronics, second because they look cool. Like it or not, most likely the commercial for such a car is probably going to be both offensive for feminists and praising machos or some sort of rebel lifestyle.

You might be tempted to say that comfort and luxury are more appreciated then the technical aspects, as most people having the money to buy these cars, are buying them just for the way they look. Comfort – in its lazy acceptance - and luxury rather go with Bentley and Chrysler, which have never inspired youth or young spirit, betting more on maturity (which is a well-played card given their audience). If you know your audience, you are not going to promote luxury details where people expect performance (valid also the other way around). It's obvious that all these four brands that I named reunite all these features – the only difference is the degree in which they mix, to make the end product attractive for their buyers.

As a good practice, a risk analysis is made to optimize the testing effort when the time constraint is weighing heavily. However a risk-based testing approach can be part of the usual test strategy from the very beginning. Knowing your user makes a qualitative risk analysis actually meaningful. Risk is assessed through probability and impact. If probability can be relatively easily measured based purely on common sense, the impact assessment requires good knowledge of your customer and his/her business. Foreseeing project and product risks facilitates early mitigation actions, which in turn lead to better end results.

Knowing your buyer (which often is the end user), involves knowing the culture of your customers. This allows you to come up with better products or at least to not offend the consumers. It prevents you from planning to hit the market with white wedding dresses in East-Asian countries. This knowledge can help you understand sentences like "The Sun in Japan is red".

Ignoring this aspect was an expensive lesson for a world-wide famous fast food chain, that had to withdraw from Bolivia in 2002, because people could not be tricked into buying junk food, not even after 14 years of effort in this direction. Sometimes I read in the newspapers sales statistics at country-level for a continent and constantly someone has to show surprise that banana sales are dropping in one country, or coffee sales are boosting in another one – well maybe the first ones just don't like bananas. However, increased coffee sales could be a side effect of expanding offices and intellectual work in detriment of manual labor or simply a reduction in taxes.

Another example of minimum interest for American vs European culture comes from an elevators' manufacturer. We are all aware of common differences between these areas, like the metric system vs imperial system, Celsius vs Fahrenheit, dot vs comma as a decimal separator, AC 110 voltage vs 220. Well, there is one other thing to add to this list and that is the way we number the floors. In Europe, ground floor is zero, first floor is 1, basement is -1 (minus one). In US, ground floor is 1, first floor is 2, basement is B. This manufacturer installed elevators in a shopping mall, in a European capital. Everyone was confused as to why should they press 2, in order to reach the first floor?

All that would've taken them to fix this was a minimum effort to replace the labels on the buttons. This also shows that none of the decision factors actually used the elevators before accepting the work delivered. It seems appropriate to mention here that if you tell someone in US that you want to have a meeting during the first day of the week, that would mean on Sunday, while in Europe that would be Monday.

Regarding the voltage part, it's worth sharing a funny story. I shave my head and as many other people doing the same, I own a hair shaver. This machine travels with me whenever I travel for more than one week. On its charger it says it supports both 110V / 220V. When I see this note on a device, I expect that machine to perform well, both at 110V and 220V. Being sold in Europe mainly, it performed absolutely fine at 220V. The moment I plugged it in, somewhere in the north west of Costa Rica, trying to cut my hair, it got stuck after one second (I prefer to express the time it took to start annoying me rather than the length of cut hair in cm vs inch). The difference in speed was the one between a whisk and a blender. The blades were moving in slow motion. It took a lot, a lot of patience to get the job done...

An editor released a numismatic collection including in each issue, two currencies (either paper money or coins; sometimes a mix) along with a printed magazine / brochure. As a numismatic collector, I've bought a number of issues, and then decided to stop because it was not worth it. In order to have constant supplies of "inserts", the editor was including mostly recent money (still in circulation or freshly withdrawn), hence not rare, not old and with no sign of usage. The bet was on diversity. For a collector, what matters more than diversity is rarity. The bigger problem was the quality of the information inside the magazine which was very poor, more than 50% of the content being over-sized images and the text simply describing the images, as if written by Captain Obvious.

So why did this editor try this approach? Mainly because it had a history of supplying full collections for different hobbies, but in a manner that could catch on mostly for amateurs. Why does it fail? First, because there is no one universal recipe that you can just apply for any type of hobby. Second, because even amateurs don't like to hear that what they have is purely common and with no value in the collectors' world.

Moving into a more common area, what does an end user want from an airplane or a flight? Cheap tickets and comfort. Of course any free stuff and good meal, during the flight, is always welcome. The key to have an airplane-product that would win is to optimize consumption. Not talking about reducing the carbon footprint, which would be a collateral benefit and good publicity. Optimized fuel consumption can reduce the ticket price, particularly for long-haul flights (typically taking more than six hours of non-stop flying) which are the most expensive. For proximity destinations, this improvement would not be visible since a large part of the price might be the airport taxes. Gaining such an advantage for long flights would win customers, preserve the profit and give a serious hit to competitors.

Psychology plays an important part in getting to know your user. Older persons tend to be conservative. If the majority of your buyers is in that market segment, they will not embrace change. Young people get bored easily and they are attracted to 'new' like a moth to a light bulb. Delivering the same recipe for a young audience, over and over, no matter what the demographic trend is, will shrink your public. You might say that fast foods with golden arches or colonel logos are the living proof against this statement. Actually, they're not, because fast foods don't focus solely on the younger crowd and yet every year they have new products on sale. Fashion companies – some of them – are focused on a younger age group and they know better than anyone what 'yesterday's news' means. Consumer psychology did not evolve as a branch of psychology for no reason.

Sex shops learned that their buyers might be open-minded, but they don't want to be open with everyone, so privacy and discretion play an important part, since most of their products are as intimate as they could get. To reach new clients, particularly those that might be embarrassed to enter a shop, they had to be active online. One thing solved, but the next issues were: what will it say on the bank account statement once an order is placed and how will the package look like? That's how they ended up billing 'entertainment (goods)' (wink, wink) and sending out brown common boxes, since no one wants this type of delivery with Christmas lights around it, especially if delivered at the office.

Cheap doesn't make a profit. Since they are accessible, cheap products tend to invade the market and potentially fulfill completely the request on the market. Unless they need constant replacement before they have satisfied completely the request, it's a matter of time until they stop selling. Cheap products need sales volume to generate a profit and if the sales volume drops, they can't make a profit anymore. Furthermore, these products easily find competition, since the production cost is easy to cover.

I think the story of the 'kill switch' or 'e-stop' fits well into this chapter. The kill switch is that big red knob you could see next to any heavy or industrial equipment able to squash a man to death or cause serious injuries. It is usually placed on a contrasting background – most likely yellow or white. Usually the button is also marked with the inscription 'emergency stop' or just 'STOP'. So it would be a fair question to ask: why is it made so disproportionate? Why didn't they make an ordinary red button saying stop?

There are two major reasons behind it: first one, no matter how accustomed people may be to operating a machine, panic can have strange effects, so the right button to push has to be visible from a mile away. Second, around heavy industrial equipment there might be a lot of illiterate people – don't think of your level of knowledge or the people you are accustomed to. Sometimes heavy equipment is sold to countries were education may be a luxury. Those people have to be able to react based on a very simple representation in their minds of the button to push.

PEST analysis is particularly recommended if you're planning to sell ice to the Eskimo. A complete analysis taking into account political factors, economic influences, social criteria and technological context, can help you understand when is the right moment to release or market a product, even withdraw from a market. Political factors can mean even laws in the making – offering let's say subsidy for solar panels. Economic factors can tell you the right price, based on purchasing power while technological aspects may discourage you from promoting video streaming in a country where a good Internet connection speed might be considered expensive. Social component can give you valuable information about how to promote your items, while the technological part might tell you for example that it's pointless to put effort into marketing GPS devices in a country for which there are no GPS maps available yet.

Disney Interactive lost considerable sums of money by targeting their games on the segment of 10+ years, in the Disney-all-family-entertainment manner (where in fact the center of attention is the kid). The gap with this approach was that on most markets, the core buying audience consisted of adults and teenagers (close to their last teen years) and here's the cherry on top: they are buying violent titles like first/third person shooters. Having this sort of knowledge can make you weight twice the return on investment and reconsider what do you actually bring so unique and addictive that can change the market trend.

In order to find out what your user thinks about your product, here's an idea from the author of "The Secrets of Consulting", mister Gerald (Jerry) Weinberg : "Say what you like in the present in order to find out what you want to change." The problem with live surveys, on the street or in a shop, done by real people is that the interviewers are sometimes underpaid students or better said 'unqualified personnel' that in order to speed up their job tend to suggest answers (by asking leading or closed questions). Sometimes they do this also to avoid awkward silence that falls after asking open questions. The opposite would be Internet surveys, having the disadvantage that you get to reach only a certain category of buyers that uses this communication channel and if you're selling cleaning products, what are the odds of someone visiting you site to find out *more!* about cleaning... There are of course options, but the focus of this chapter is on finding out what is important for your user.

The more expensive the product, the more dramatic a defect becomes (especially on new products). This is a general rule, although the impact is logically higher for someone with a lower income. Once you've paid 400 euros for a new smartphone, you don't want to see the slightest defect on it.

The things described in here are not project friendly because they can't be quantified exactly. However the investment in getting to know your user is minimal, once you know what you have to do or just follow some basic guidelines.

That is because in the same manner we want to observe species in their own environment, to see their natural behavior, the same thing we can do with consumers, by observing silently how they choose a product, how they interact with it, to what features they pay attention and so on.

If you setup this study in a non-standard showroom, placing let's say functional mobile phones on stands across a shopping mall and setup also a hidden camera, you might observe the following (these are just examples of what you may find and not the results of an actual study): that most persons drawn to study the product are teenagers; that if you place a phone with white case next to one with silver case, women will be more attracted to the white one; the fact that if you place several functional apps on the phone (e-book reader, HD video content, music content), users might check the camera first, then the sound or perhaps the data connection.

All these aspects can make your business friends with quality and buyers. The original concept of all these theories that have evolved to the level of science, was a very simple one called: 'giving a damn'. There are plenty of ways however to drive away customers which we'll see in the next chapter.

The natural enemies of quality

Behavior specific to corporations listed on the stock market and boards of directors

These are on the top of my list for several reasons. Stocks are worth only if their value is growing. Usually listing the company on the stock market and having a board of directors go hand in hand. A justified measure since you wouldn't want a kamikaze guy sitting on billions of dollars to make bold decision depending on his instinct, or horoscope, or perhaps a dream he had the night before, betting with other people's money. Simply maintaining the stock at the current value is not enough, because nobody would make a profit from selling or buying these stocks. So their value has to grow. But these boards refuse to understand a basic principle of physics that applies to economics: Newton's law of gravity which simplified (a lot) for dummies would say: what goes up must come down. That applies wherever there is gravity around, so that's pretty much every place on Earth.

Independent of a company's behavior, the economy has periods of contraction and expansion. As long as performance is constant or improving, any natural decrease will be followed by an increase, so putting that into a graph, good *realistic* performance ends up looking like a Gauss bell. Always. So if we would use a more accurate term and call this bell curve, a Gaussian distribution or normal distribution, the higher the standard deviation from the mean during increase, postponing the downfall, we perceive it as good performance. Even if in some person's wet dreams, it would be just constant growth, eliminating all competition, there are laws against monopoly (we're not discussing natural monopolies).

So this being said, what's the connection with quality? Well, no board will accept a strategy based simply on organic growth and steady profit. The Board's efficiency is translated into maximizing profit figures by reducing time to market, cutting costs (cost reduction) and reducing the actual figures of inevitable spending.

Think about the difference between one or two persons (even a whole family) owning the company versus having it fully exposed to the (stock) market. When the company is not listed, there's time for philosophy, there is time for saying 'we are making only the best products and we're not releasing this one until the quality is gold level'. It allows one person to say: 'I may lose a hundred thousand euros, but I'll get something good on those shelves, so you may have that one more month to get the product in a better shape, because it's my reputation at stake'.

Contrary to popular beliefs, not all guys owning big companies are greedy bastards and quite a lot of them have a good self-esteem and personality. A personality that makes them identify with the product itself, so if a lousy product gets to the customer, they feel lousy. But when the company goes public, shares value has to rise – that's the only thing that matters.

Attention to scandals, image blows, reputation, all that is connected indirectly to shares. If the image is bad, if the company looks corrupt or discriminating, people or other companies avoid associating with it, because it will stain their appearance. Why does a refreshing drinks producer care about advertising as a 'green', 'eco' producer? Why do they promote equality, friendship and all positive principles? Because their target consumer is in fact any person who can afford the price of a juice/soda can. Why don't we see weapon manufacturers advertising about how green and non-discriminating they are? Because they are dealing destruction (labeled with hypocrisy as maintaining 'peace and safety') and when dealing death, your buyers couldn't care less about the environment. In fact, they are honest about one thing: weapons don't discriminate.

The end reason for companies 'behaving nicely' is that a consistent number of buyers could refuse buying their products or even boycott the company, while NGOs could start legal actions. Involvement in lawsuits, especially losing them, can severely affect future contracts (government contracts or simply partnerships). All this means a drop down in sales, that is a drop down in shares value. This also gives you the key to knowing which buttons to push if you really want good quality in such a company. Play these cards for COTS (commercial off the shelf) products: public image, discrimination (if accessibility testing would apply), reputation.

Real-life example: for political and budgeting reasons, a company tried to push its product out on the market, although the end report from QA said it's medium to high risk and the overall feedback was negative. The product faced a lot of problems in production, so many issues in fact, that a lot of users gave up using it. This led to financial losses and a small number of managers getting fired.

What does reducing time to market lead when it comes to quality? Ever since the waterfall model was eliminated or started wearing the stigma of inefficiency, quality has been evaluated in parallel with development, except for one inevitable phase at the end, where development stops and there is still a significant amount of testing to be done. Well, what happens then? In some companies, pressure moves onto the test team, wanting them to finish faster. This phase is perceived as 'too much time wasted', because the product is 'ready' – where 'ready' can range from 'good to sell' to 'time bomb', filled with last minute or quite well-known defects. Shortening last testing phases is a form of cost reduction – especially as the budget is not doing very well towards the end of a development project.

Second form of impacting quality through cost reduction is hiring less skilled personnel, on the grounds that two or three juniors will learn from one senior. Of course, a senior could train and mentor juniors, but two juniors will never equal a senior. How come? Well, if you're building rockets and bring in two persons that dream of designing rockets versus one person that actually built a rocket, you'll end up with two persons still dreaming to build a rocket and an over-worked third person. Two wrongs don't make a right.

Professionals, good professionals, are never cheap. Even if on paper, the paychecks of two juniors would equal a senior, this doesn't make up for the knowledge and let's face it, neither for the errors that juniors are prone to make in order to learn and gain experience. We could brag further about motivation, good HR selection process which can help you pick exquisite juniors. Yes, these are all real. So is winning the lottery, yet I don't see people betting their retirement funds on a raffle ticket.

Here steps in the third enemy:

Processes.

In big companies processes are a necessity. It's hard to control a big crowd without them. Therefore, I will rephrase this: the third enemy is: the perception of processes. Mark my words: *Quality comes from people, not from processes*. Simply documenting and archiving documents will not make a difference. It takes minimum six months for a person to actually know how things work on an IT project. It takes more for that person to actually be proficient and bring in value. Preserving your personnel preserves the existing quality. If you don't have any real quality right now, expecting different results while using the same persons, so the same input, is not very clever. I don't want to turn this into a speech that's not appropriate for the purpose of this book so I stop here with this idea. Remember: The value of a company is in its people and assets. You may be able to track the assets, but you don't know when those people who were (actually) making a difference left.

Mixing suppliers in a team

... is never a good idea. In fact having too many suppliers, especially in a project is not a good idea. Why? Because the new motto of the team will be "it's not my job to do it". In more professional teams I noticed however that since each supplier has in the end its own budget and profit target, plus a set of deliverables to get done, "man-hours" become a trade coin. You want some other team or person to help you do your job or simply understand something? You can forget the word "collaboration". Either you have some "hours" to give-away or you don't.

The more suppliers you have or even freelancers, as subcontractors, the harder it is to establish back-to-back agreements (don't interpret this in its legal meaning, but rather in its wider concept). More parties involved means more individual interests and stakeholders' management is not going to help here. Mixing more parties in the same team is purely suicidal and nothing good will come out of it. This is the case when inside a department which got outsourced, you decide to keep: one or two employees of your own, one collaborator from the first contractor (because he/she was good) and the rest are employees from the current contractor, plus, why not, an independent worker. Simply asking them to perform as a team will not work.

Our next foe, brought into the light:

Bad theories taken as axioms

...like "Parkinson squeeze", "the legend of 'quality is 30% of development time' " or the urban myth of test automation. Like it or not there is a thinking pattern in project management and management in general that says the manager does not need to be in touch with the technical aspects: he is just managing. Unfortunately we encounter this type of 'pure managers' born from the foam of the sea, superior to mortal creatures and knowing all about nothing, all over the place.

This is the kind of people that think cutting off spending on single use plastic cups is a big achievement; the kind that think all cost cuts made in spreadsheets, actually work in real life; the kind that when seeing a graph about product quality are able to ask only 'why did you pick the red color for that line? Red scares me'. These are the ones that tell you a business travel fits perfectly within company policy and nothing unexpected ever happens. The kinds that actually believes you can pay the taxi by credit card in an underdeveloped country or that ATMs are everywhere and safe to use. This class of managers, which I wish to you never to encounter in your professional activity, take for granted all theories received in expensive courses without questioning. If you are a manager and reading this book, most likely you are not part of this class.

Let's have a look as three examples.

First one – the Parkinson squeeze, which fits better among 'Murphy's laws'. Simply put this law states that *'work tends to expand to fill all available time'*, meaning that if you have one week to do a task, you'll do it in a week. If you have two days, you'll do it in two days. This may apply to students, children, lazy workers, but never to professionals. If an experienced staff tell you it takes a week to finish an activity, granting them four days will not speed up things. Forcing this rule might prove if your staff is indeed professional, just as long as you don't rely your budget on applying the Parkinson squeeze.

Second example: *"testing takes roughly 30% of development time"* – often encountered in software development. Well, completely non-technical persons may not get this, so let's try to put in simple terms. Do you know those messages popping on the screen when you complete an online form saying something like *'Full name is mandatory!'*. For a developer to create an alert similar to that, in a scripting language, it can take less than a minute, especially if similar alerts are present for other fields. Yet, writing a test specifically to test a newly added message, just typing the test (since we have to document our work) can take a few minutes, not to mention running it and reporting a defect, if one is found, plus retesting it after it's fixed. So how is that 30% of the development time?

Modifying access rights for a user role, as part of a change request, which can be as simple as removing a check mark or commenting some code, usually means a number of hours of test execution. Then add also test case design and preparing some test data, if needed. How is that 30% of the development time?

Third example: the myth of test automation. This type of testing has its benefits, it's mandatory in certain situations, long term projects, there's no point insisting in this direction, as I believe we can agree on this aspect. But their creation requires an extensive amount of work and sometimes of rework using expensive tools and well trained staff. The myth is that if you have automated tests, you just press a button and you'll get the response: 'All works!' or 'This specific thing is not working', as if no one has to analyze the end results. As efficient as they may be in the context, automated tests are never cheap and are not always worth implementing.

Faking

...is the next enemy. Calling defects, 'features'. The use of euphemisms to mask a problem. Even the 'bug' term is an example, because a crash is still a bug – so we would call a major f*** up, with the name of a little crawler, that is unpleasant but doesn't really kill anyone. Faking. Pretending you do not have a problem when your experienced personnel is leaving like a swarm leaving the hive to relocate. I witnessed an institution losing 50% of its personnel in less than one year; employees that left by resignation. Nobody asked them to resign, there was no re-organization plan behind it or a major player showing up into the market sucking off resources like a vacuum cleaner. No such thing. In this context, this institution thought it didn't have a problem and it's just a "transformation". Well, at least in this statement there was a gram of honesty – nothing is lost, everything transforms.

Corporations are often the patient who doesn't admit he has a problem: because of the stock market. When the CEO of a major German group publicly announced that realistically speaking the group cannot meet the targets set for that year, in a context of economic crisis, he was fired the next day.

A country manager working for a Japanese electronic producer in EMEA division, made a similar statement, but this time internally, indicating there will not be any profit that year due to the crisis, but hopefully no loss either, having the financial reports and predictions at hand. This honest guy was severely admonished by the board for not targeting a profit, even if figures showed it couldn't be achieved. Corporations sometimes set SMA*T objectives, without the R, cause they are anything but realistic. Usually the timeline is the one which is not realistic: they push to release, they push to save budget, they cut down on quality. The buyers get tricked once, they might buy again, still remembering good past experiences (if any), but the third time, they will fall for idiots – shame on them.

The first step to solving a problem is admitting you have a problem

In a project that I once worked on, the client assigned some unskilled staff to create a technical documentation. They did what they could and the deliverable coming out of their hands matched the expectations. The negative ones. This document was supposed to describe the integration between the system being created (a web based application) and a third party system (another web based system), already in production. During this documentation, no one had a face to face meeting with the third party to ask them technical questions, since they were the ones that knew best how their software worked. Problems were increasing in number and the two systems did not seem to match in any way. Finally the unavoidable meeting was established and those guys ('the other team') sent their documentation to serve as basis of the discussion. Comparing what our client had with what the others sent, I stated that they had nothing in common. These were the facts. A rework started, a lot of things changed, all with an important budget cost, affecting everyone, because management refused to acknowledge earlier, what everyone was saying: something was fishy about that documentation.

Auditors

Please pay attention: auditors, not audits. Generally speaking, an audit – especially a financial one – is well justified. The problem is with the internal audits, assessing compliance to processes and regulations. A good process prevents you from 'forgetting': it gives you checklists, templates, artifacts. Although it's more than a guideline, since it's mandatory, its essence is one of guidance, it's meant to help you do your job better. Naturally, a process allows exceptions as not everything can be foreseen.

Problems start when auditors that don't understand the nature or context of a process, verify the 'ad literam' compliance and force teams to blindly follow a set of rules that they can't even explain. These persons can generate a tremendous overhead of work, by their simple presence, so instead of having people focus on the productive work, they start spending hours on compliance with internal regulations, hours billed from a budget that is not magically growing. The best way to preserve a stupid rule is to let an idiot guard it and make it his/her sole activity.

Rules that don't make sense are covering-up past (unfortunate) events. Lawsuits or losses. Why do you think there is a warning on your coffee cup letting you know it's a HOT drink? Justified bureaucracy is covering such events that took place so long ago in the past that nobody remembers how it all started.

Stupidity should not be encouraged or tolerated. I am fully aware there is a history of law suits where consumers have won, complaining they were not explicitly warned that "sauna is hot", "fire burns", "looking at a screen for hours may damage your sight", but trying to make a product idiot-proof is a wrong approach. Common law shows that going against the grain in this aspect can cost you a few million dollars. So what's the solution then? Disclaimers and user acceptance of terms and conditions are the only one which I would recommend, unless we want to start making "hammers with hand detection" or "chainsaws with face detection".

Chaos

It's always easier to succeed in a place where disaster already hit, because worst case, you might be remembered as someone who tried but it was too late. If you do succeed, you'll be a hero. Arriving in a place where there's already chaos, but not acknowledged, you might be perceived as a contributor to the bad situation.

Ignorance

Gerald Weinberg – an author and consultant that I appreciate, issued in one of his books the Titanic Law: *"When you think disaster is impossible, you're heading for an unthinkable disaster"*. That is why recovery plans are not in place until something really bad happens. I used to work a long time ago on a project that was basically transforming meter readings into bills (commonly called meter-to-cash system). A downtime of one hour of this system could cost the client hundreds of thousands of Euros – in other words: a lot. When I realized this, I proposed to start some basic recovery testing (different scenarios), to see how the system recovers lost information, how it establishes connectivity among elements and so on. The idea was immediately dismissed by the project management, which took into account only the existing budget instead of trying to sell these additional services.

Nothing happened for six months, until the customer who knew he couldn't afford to have a single-point-of-failure, reached the same conclusion and demanded these tests and a recovery plan in place. This is one example, but ignorance has many faces. Therefore, if you create a new product and start by assuming all clients exposed to it will buy it like crazy, you're business is going to collapse faster than a tree surrounded by wood choppers.

My advice is: **always listen to your experienced staff**. If you can afford to ignore them, you're not paying them enough. These people may seem conservative, rigid. It may well be the case. Resistance to change is natural in humans. Bright people don't oppose blindly to change, but they don't embrace it blindly either. They have to question everything. A proof-of-concept is what convinces them.

The role of consultants is not... magic. Their most important attribute is that they can have an objective view without caring about the impact of speaking out loud, at a personal level. Sometimes corrective measures can be radical, shutting down projects, firing people, admitting failure. Of course, consultants can't impose these decisions, taking ownership of a business that's not theirs. I've been part of a company where during the financial crisis, firing decisions were made without any business reasoning, putting first criteria for keeping people like 'who's got more kids': a purely socialist decision taken to the limit of absurdity up to a point where managers in their right mind had to intervene to correct the dismissal list.

External consultants have a role similar to a shrink or an honest friend: they tell you what they see, not what you want to hear. The key to benefit from this is not to put any filters or 'modulators' in between, that will interpret their message. A manager who is hiring external consultants needs to have direct contact with them and take into account what they say, especially when it comes to quality consultants.

Assigning unqualified personnel to do a tester's job

Why not use just business users (specialized end clients) for testing since in the end they are the ones who are going to use the end product? Because there's a high chance that they will exercise only common use scenarios, common paths and are not aware of many technical requirements. Why not use business analysts? Because some tasks may be too technical for them, so they can't cover that properly and they will not be able to see everything through the eyes of a user, either – so this is one of the worse solutions. Functional testers need to be half technical, half business oriented, while wherever there is a particular need for validating fully technical aspects, specialized personnel can be brought in: security testers, performance testers, etc. The bigger advantage of using dedicated staff is the analytical thinking combined with testing techniques.

Let's pick an extraordinary simple example. You have a field in a webpage called 'Tell us your opinion!' where you can enter free text and a submit button. A business user would just type in some text and press Submit. A business analyst would type in some text with punctuation or special characters, maybe both and submit it.

A tester would check: does it support regular text, numbers, special characters from different languages and punctuation marks? What's the limit of this field in the database - is it defined as string, limited to 255 chars? What if I enter 256 chars and press submit? How about exactly 255 chars? What if I leave it empty and try to submit it? I can't type more than 255 chars, ok, but what if I copy paste some text, will it trim it? Does the text get stored correctly in the database? What if I try some html injection (cross-site scripting) or sql injection? What if I close the page without submitting: does the text get cached? If I can enter more than 255 chars, is the trimming done on the client side, or on the server? What if I enter only blank spaces? What if I double click the submit button or click on it like crazy? What if I intercept the message after trimming on client side and edit it before reaching the server? Trust me, this list could go on, depending on the goal of the test.

User fault reports

A common user will give you inaccurate bug reports and present false expectations. Relying on the accuracy of a report from a real user is pointless. Not because they lie (although everybody lies), but because they omit things. So this could be a philosophical question: is someone lying if he's convinced he's telling the truth? This is what is causing that stupid exchange of phrases when calling customer support about your computer not working and they first ask you if it's plugged in and turned on.

During my work days as a support officer (I love this title, it's full of 'officers' in banking and IT and a lot of 'cockpits' as well) for an antivirus producer, when users were reporting some particular errors we had to ask if they have another antivirus installed. The answer was usually 'no' and whenever we got a particular error message, we knew for sure it's most likely a 'yes'. If the user agreed, he was being asked to run a particular tool on his computer and send us the report. Guess what, we were always finding traces of another antivirus. When facing the customer with this evidence, answers were ranging from *'Oh, that one!'* to *'I even forgot I installed it'*, both of them honest answers. You just have to address the right questions with diplomacy, but don't believe blindly whatever users a reporting.

The connection between user reports and quality is that all reports need to be analyzed first, reproduced in a clean environment and only when the problem is completely isolated and reproducible, a root cause analysis can be started. There is a seed of truth in every user error report, but it's covered in layers of irrelevant information. Inexperienced (project) managers like to keep an eye on defects reported in production and on server error logs, jumping to conclusion before they understand what is really going on. It is QA's job, to analyze these reports at a bigger scale, see where the problem is and make recommendations in collaboration with the BA and development team, for future enhancements or hot fixes. This depends on cost of fixing (given by effort) and impact if not fixed immediately, available workarounds, etc.

Cost-benefit analysis

...is in practice – most of the time - a cost analysis ignoring the benefits, because trying to quantify benefits is something subjective. Example: how do you quantify the negative effects of inaction? *"The price of inaction is far greater than the cost of making a mistake"* – quote attributed to Meister Eckhart. Not releasing a product on time can mean losing momentum – everybody knows that. What about that extra bit of testing before releasing it? Yes, it does cost, like anything else in the world. Is it worth or can we just skip it? The test team might know the answer to that.

Assessing the marginal utility when you're not working with factory products is something highly subjective. When the components used in production are 90% intellectual capital it's hard to say what benefit will bring you an extra day of work – not to mention research where this is nearly impossible. So who should do such an analysis? People involved hands-on in producing the product. Who's actually doing it? People looking at spreadsheets and tables.

Do you know why finance software is so damn ugly and why finance software made on demand will never be as good as COTS? Because financial institutions always want to cut on the development cost. Because they tend to use blue, black and white and SAP-alike screens to keep a stiff appearance as if a touch of colour would make people go crazy and lose focus of what they are doing (while in fact color blue psychologically creates a state of depression, but also induces logic and serenity; black inspires security, but also oppression; white creates a mindset of purity, but also coldness). Because financial institutions use old technologies, so old that the first schools studying those coding languages were created around monasteries and the first handbooks were written on papyrus, with runic symbols. Because like it or not, you have to use it and account statements are often a usability nightmare. So once again: do you know why finance software is so damn ugly? Because they deserve it.

Last, but not least, companies (or countries for what it matters) that don't invest in education (training) and research (new products), yet expecting to rise, will be just temporary passengers in a train called 'Quality'. They have no ground for a healthy development and expansion.

The solution is at hand for those that want to see it: accept that profits will drop at some point – it's part of the game; hire professionals and listen to them; invest in learning because things are changing faster than ever before; call problems by their name; don't audit for the sake of doing it; be prepared and see the long term benefits of investing in quality.

State of the art

Trying to bring whatever you're doing, whatever your focus is, whatever the product, to a 'state of the art' level should make you ask yourself one simple question: can your audience appreciate it? Let's face it: haute cuisine is not for hungry people.

One could argue that once the level of quality changes, so does the target audience. Correct up to a point: it's true only if the price changes, but then also your competition changes. More important is that in an open market the offer does not drive the demand. Next question is: what sort of recognition do you expect from your audience? Money, public recognition, praise?

If you want to create something unique, artistic, you're crafting at the top of Maslow's pyramid, but if your customer is not there already or one level below, your work is not going to be appreciated. Do you have access to your audience? Do you know your buyers, your consumers, have you done your market segmentation? If they are at the basis of the pyramid, your work will not get its message to them. This would be a good way to know where to place your new product in terms of quality and in what way to promote it.

If your consumer can be placed in terms of price range at a low level and is targeting individuals with limited income, focused on savings, it means your consumer is working on covering the basic needs. So your target level would be one level higher, promoting in terms of advertising comfort, safety, things about which your consumers dare to dream and achieve. Your product is within reach.

Insurance business is working in the wrong direction on this aspect. Insurance is based on probabilities (from a mathematical perspective) and on fears (from a marketing perspective). The wrong approach is that for example a life insurance – usually not a cheap product – which in terms of price addresses the third level of the pyramid – people affording social interactions, having a social life - is creating its marketing campaign around level 2 – the safety level. I believe a much more efficient way would be to stimulate the buyer's needs of recognition or of preserving the current lifestyle, with the peace of mind such insurance would bring.

Manipulative as it may sound, this is pure psychology and the example above can be extrapolated to any other product. Most avid buyers of smartphones are teenagers and young people, living at the 3rd level in the pyramid of needs, so what message are the commercials sending? If you buy this phone, you'll have more fun and interaction (level 3 of the pyramid) or if you have this phone, you'll be the coolest in your group (recognition – level 4 of the pyramid). This would be also a good example of how the end user could be misidentified, because although I used the word 'buyer', most teenagers are not the actual persons making the payment, but they are driving the purchase. This means that if you would keep an evidence of who paid for the product, you could end up believing that most avid consumers are people in their forties or fifties.

Back to discussing quality and to sum this idea into a conclusion: the level of quality you need to achieve is the level at which you market the product.

To reach a state of the art level, you need more than simply "talented" people. Artists are people with talent (having thus a natural gift for doing specific things right) that love (or live for) what they're doing. They are willing to work overtime, out of passion, but not to be exploited.

Please note that I use this term 'artists' in admiration for technical people passionate about their work that actually make a difference in the world (creating mind-blowing things like solar highways or raising crops in the desert) and I do not include those that poetically put: 'move to the sound of music, to make the world a better place'. The latter category includes also at an extended level, all philosophers that meditate about the essence of things, instead of actually doing something.

Artists, living at the top of Maslow's pyramid have another interesting trait and that trait is the desire to change the world: for good or for bad, because a genius is a genius, while a lion let's say, can be good or bad depending on which side of the fence you are (thus a matter of perception). The difference between them and 5-star entrepreneurs is that any profit in the attempt to change the world is a collateral benefit and not a goal. Jonas Salk – world famous for successfully testing on humans the polio vaccine – a prophylactic treatment for a disease comparable in modern times with the plague, never applied for a patent.

However, a bunch of lawyers did analyze the possibility of registering the patent and making a profit on the back of those that wanted a better chance for life. Fortunately, the original concept and first vaccine tested on primates, as were the times then, was developed years before by Isabel Morgan's team – a name which most probably, you haven't heard before.

This made any patent registration impossible, and among other things, it changed the face of the world forever. As an immediate impact, it also made people paint their windows with "thank you" notes, immediately after the public announcement of Mr. Salk's success.

Nearly sixty years later after this event, we've seen a bold move in the automotive industry, where car producer Tesla decided to publicly open up its patents to those who want to contribute to electrical cars development. Of course, we can't be so gullible as to believe this is a free act of kindness from a company listed on the stock market, but it's all for the best and quite a good strategic move. We should be able to see in the next two decades if their idea paid off. This was simply one good answer to the question: "How do we speed up the development of a global platform for electrical cars, so that we can increase our sales?" The challenge nowadays is not obtaining the information, which through Internet is widely accessible, but to be able to actually use it (legally, financially and as a production capacity).

As we were discussing 'artists', there is always a sense of rebellion in artists and I am very amused whenever I see "rebels" working in corporations (these are not artists, trust me). I would love to hear more about how "rebel" these people really are, working eight hours a day at a desk in an open space or a cubicle. They can go ahead and tie bandanas to their mouse, while doing the latest online training or corporate compliance activity.

The problem with real artists is their sense of perfection and the tendency to invest extra time and resources, to make things perfect. Particularly these two resources that are never unlimited or even close to being convenient and affecting two out of three major elements of a project: time and budget.

A state of continuous improvement is the closest thing we can see in corporations or factories. As admirable as the Japanese concept of kaizen may sound – which says that every person, from factory worker to CEO, is making an effort to improve productivity and quality – this concept has a significant cultural component. As much as Japanese methods present interest and are worth studying, they can never fully apply on a Latin culture, where people value more 'enjoying life' and less the 'perfectionism' at work. Nevertheless, kudos to Japan, for making an insular state, shaken by earthquakes, able to compete with fifty United States or other continent-sized nations.

Linking perfectionism back to Maslow, one thing is sure: excellence does not come from demotivated environments. It's always difficult to keep a team motivated, so personally, I find it hard to believe when I read in the latest tech newsletters, that some company that has just fired a two-figure percentage of their employees (context in which all the remaining ones feel a sword swinging above their heads) will make a spectacular comeback and blast all competition. This would be possible, no doubt, in the presence of an inspiring leader that can literally send a message that bad times are gone, sacrifices had to be made, but 'we're up to something major'. Such persons are one in a million and often they are undermined by their sub-executives. I met only one such person in my entire life. So, simply put: starving dogs don't do tricks.

Artificial increases in share prices or profits, puffed up with all sort of budget cuts and modern slavery (tons of unpaid overtime hours) are never going to lead to innovations. While the figures in excel sheets are rising, the moral of the people is dropping and here we can close the loop with the statement I made earlier: demotivated people cannot excel.

"No software can be better organized than the team that creates it" - Noah Sussman

(more interesting details are to be found on his website: Infinite Undo – I hope any reader of this book is able to trace the website, without an explicit link; if not, please return this book)

This is an interesting theory which I consider to be true. I also feel the need to add that software becomes less human, whenever the company creating it is getting more ... organized. The more metrics, charts and reports you have lying around, the less user-friendly your software will be. This doesn't mean that excellent products bloom out of chaos. No. It's just a matter of shortening the red tape and the unnecessary bureaucracy, often masked under the name of 'organized'. Trying to enlarge a bit, in really big companies, you have dozens of people so specialized that they are doing just a tiny little bit of the work, so little that they can't see the forest because of the tree in their way. A product owner should have a global vision, but he/she can't, being suppressed by paperwork and analysis. Here is where agile comes in handy, making people...ah, forcing people, to put on different hats and understand more than just what they are doing.

Speaking about software: in software testing, the concept of 'kanban' made its way through several books and materials – again, an interesting topic to study, which with a bit of effort can be applied to abstract things. 'Kanban' was invented by Toyota and it was a means to control the production cycle, through the usage of some kanban cards, the demand driving the production. There is never an unnecessary stock and target is to have 100% defect-free goods. This is the story in a nut shell, although the mighty Internet would be happy to tell you more about it, if you're interested. As much as it is worth studying and remembering the 'just-in-time' approach which is encountered also in agile development, this is a concept that originated in supermarkets, was applied in plants and relies a lot on 'kanban cards'. So: nothing in common with software, unless we want to debate philosophy. I've heard about companies trying to apply this in software development, but ended up creating a hybrid, a mix of notions that we could say it originated in this Japanese concept. However if someone would like to share with me how this method was applied in software testing, in Europe, I would be very interested in learning about it.

Sticking around the same geographical area, it is worth mentioning in this chapter an expression inspired from the Japanese 'hansei' (a concept that implies admitting your mistakes and striving not to repeat them) and that is:

"No problem is a problem" – a phrase more fit in a European context.

It is nearly impossible, at the end of a long term effort, involving a team and a process, to say that everything went flawless and no lesson was learned because everything was perfect. That is simply a lie. Even if it would be true, ways of improving could still be noticed, taking things from "good" to "better".

As a side note, the reason why 'hansei' concept cannot gain significant capital in many cultures is because the phrases 'nobody's perfect' or 'it's human to make mistakes' are way more popular.

State of the art cannot be achieved in a short time. It simply takes time. It is a lasting effort built day by day, accumulating at a steady pace. That means patience, investment and accepting the fact that great, lasting things are not built over night.

Quality never comes in big quantities. 'Limited edition' of deluxe products has nothing to do with this principle. First edition is meant to win the market, but usually first bunch of the second edition is the one that's best. The 'limited edition' label is just a marketing artifact made to earn a quick income, because the edition is so limited, that you have to buy NOW, if you want to be one of the few 'special' owners of this product (and of course that will make you a more unique person, etc, etc).

Prototyping is mandatory if you want to implement something that is going to be ahead of everything. Most people are familiar with prototypes from car manufacturers, but prototypes can be used in nearly anything else. Big companies prototyped their stores, to examine the user experience; ships' interiors are usually prototyped by simulating a whole experience on board of a ship.

A prototype (of anything) answers in essence three questions: how it will look, feel and work like? Prototyping creates a tangible artifact that opens your eyes to possibilities, options, corrections and change. Here is another trick: when there is no commercial solution available, lead users (which are not typical users and are also more than early adopters) already have a custom solution. That means your first prototype might already be there, you just have to look for it. And guess what: these lead users are working at the last level of the pyramid.

Quality of life

You transmit quality. What you're showing is in fact the indirect quality of other elements in your life, like food, fitness, social environment, relationships, etc. If the ingredients are bad, most likely the end product will be bad. Unfortunately it doesn't work 100% the other way around: eating good quality food, staying fit, living in a good environment cannot guarantee you will have the ultimate success. It's not a guarantee, but often this is the way things go.

Sometimes the product that you're selling could be... You. Maybe you are a consultant, a freelancer or a service provider and you are trying to sell your own work. You could be a speaker at a conference or a presentation, where even if the material is not 100% your genuine work, you are still selling yourself.

Listeners will always analyze speakers by the way they look, at a subconscious level. Even if what you are saying makes perfect sense from a technical point of view, if you show up on the "stage" (just to use a generic term) with dark circles under your eyes, pale looking, ready to fall apart, the subconscious will tell the listener: "look I don't know what this guy did, but if he practiced what he preached and he ended up like this, you mustn't follow this path or else you will end up the same". The reason why commonly teenagers are saying "I don't wanna end up like a geek" is not that they don't want to end up with a brilliant mind. No, they don't want to end up with the other features commonly associated with geeks, particularly lack of social life which is paramount for teenagers. Harsh truth.

So, if you want to improve your own life, you can start with fitness and quality of food, since these are things that are at hand to control. Fact is -like it or not – that a healthier looking person does have a better chance of getting a good job with a good salary. I said 'healthier looking', not 'better looking', because the subconscious of a recruiter or an interviewer will have a word to say. So you can start any time by adjusting what you eat – cutting down on the fat, meat, fried stuff, reading more on the ingredient tags before you buy, because in the end you will spend the same amount of money on food, but replacing fat with vegetables and fruits, might just make you feel better. Speaking of meat consumption, people should ask themselves one common sense question: how come in some cases meat has become cheaper than vegetables? Or in other words: how can the end product be cheaper than the cost of production? Well, because of steroids and hormones.

This book isn't about nutrition, but it's worth analyzing from the quality perspective how tomatoes in supermarkets lost their taste. We can all agree that for a farmer, having pests compromising the crop can be a financial fatality. We can also agree that the color red is highly defining for a tomato. With these mindsets, researchers have developed in the past decades, breeds of tomatoes resistant to pests, but still looking bloody, fiery red. This artificial selection process made through biological engineering came with a price: the end product lost its taste. Joking a bit on the side, maybe this happened because researchers got sick of tasting tomatoes after a while or they were simply too afraid to taste what they injected at some point. I don't know if I'm more of a gourmet than other people, but I highly appreciate the taste of food.

The chain of events is simple: farmers who are generally poor quickly adopted a breed that would allow them to produce and sell more; the supermarkets also valued more any vegetables resistant to transport and depositing, so in the end you couldn't find any alternative to these thick-as-rubber vegetables. More shocking is that we're having new generations that grow with the thought that a tomato is red and tasteless and sooner or later the demand will drop. This story shows how the paid / delivering entity (researchers) build the product right (according to farmer's specs, having a purely economic vision), but not the right product – because they forgot what matters to the end-user (the actual consumer).

Back to improving the quality of life: fitness doesn't have to be an extra cost. There are hundreds of exercises you can do at home, in your free time with a simple pair of dumbbells or your own body weight. People often complain about time – until you make it a priority, you may never have the time for exercising. This is true about any activity once you're past your university days – you will never find the time for learning a language or a new skill, unless you decide it's a priority. If you are finding the time to sit on the couch and eat chips, you definitely have the time to exercise. I'm not going to preach more on this subject because I don't want to impose some subjective beliefs to anyone. If you want to have different results, you need to change the ingredients OR the way you mix them.

You can distinguish quality in your everyday life, on every purchase that you make. Expensive is not by default "good". Some years ago, the "M." bracelets (I'm avoiding to give the real name) branded with the name of an ordinary TV anchor hit the market. They were made of an ordinary red wire, with a golden microscopic trinket advertised as lucky charm. Of course, fashion is a market with its own particularities, and most avid buyers seldom impress by their IQ. The bracelet I was talking about was being sold at a price of about 50 euros, with a production cost that didn't exceed 5 euros and with a bit of search one could've found better models in Chinatown. You have to analyze what you're buying.

Buying products that are not worth the money, from the quality perspective are a direct hit to your budget, cutting down on your possibilities of doing something that you like and costs money. "Expensive" is indirectly related to quality, not directly.

Why is a diamond so expensive? Most of the diamonds come from exploited African countries, going through civil wars and often come from an illegal market. They were formed through a natural process out of carbon. The industrial properties of a diamond are undisputable, yet people rarely buy them for their level on the Mohs scale. So how come a mineral formed naturally and picked up by exploited prisoners ends up being sold so expensive on the developed markets? Because it shines beautifully and because some time ago the phrase 'a diamond is a girl's best friend' was marketed. It could've been any other material hard to get. Just think about it.

Quality takes time. You've heard that before, but in a slightly different context. People are reluctant to accepting this, although they accept that 'Rome was not built in a day'. There is a constant push these days for delivering everything fast, for reducing production cost. You can see trainings and seminars on reducing 'time-to-market'. Building quality in a product is like building a house: first you need a foundation and no matter how many workers you bring in, once the concrete is poured it needs a number of days to become solid. Only artificial results appear overnight and people that go the gym regularly, know this.

This is why I never trust version 1.0 of a product. Many people learned this and so did the companies. Therefore nowadays sellers are marketing from the very start version 2.3, version 4.0 and so on. This is of course illegal if anyone admits it, but it's easy to avoid the law by pretending 1.0 was a beta release, 2.0 given to a limited number of users, etc, etc. So, let me rephrase what I just said: I don't trust the first marketed version of a product, no matter what number it bears.

There is a limited amount of quality you can include in mass products. Mass products have nothing to do with the distribution or manufacturing process (if you're thinking assembly lines). Mass products have two characteristics: they are cheap (or at least very accessible in their price range) and easy to find in relatively large quantities. Remember the saying: 'it's cheaper for a reason'. To make the end product cheap you need cheap labor, cheap materials and/or give up a large amount of the profit and hope you will sell enough. Skilled laborers are not cheap and cheap materials aren't that good. Now once you've accepted this, think about the fact that food in its general, supermarket acceptance, is a mass product meant to serve 7 billion people. Or if we exclude the starving ones, at least 5 billion people. How do you feel now about what's in your plate?

Looking at the world as it is in 2014, in the context discussed in this chapter, the one thing that could change the face of the Earth, would be electrical powered engines used in all vehicles, ultimately sun-powered. This is not even a move that would affect car producers – of course they would have to change the models they produce, but it would not affect the number of units sold or their profit. What would such a change do? It would lower ALL prices, as the total price of any good that we buy, includes transport. It would put a stop to a number of wars fought in the name of democracy and human rights, while no one cares what happens in a country without resources. It would not lower the profit of existing companies, it would lower their production and distribution cost.

This is the kind of change that actually brings quality in life. Yet, there is one price to pay: the "half-sunset" of the oil industry. As long as plastic and derivatives are still in use, it will not disappear. It's a small price to pay for the whole mankind, as long as decision factors would ignore financial incentives (so called 'donations') coming from this industry.

Ironically, the Arab states that are world renowned for their oil resources and oil business, would not have a single problem benefiting from a sun-powered energy source, which I consider karma's way of compensating for desert lands.

There's a tremendous difference between rich and wealthy. You can become rich overnight, hitting the lottery jackpot. You become wealthy after decades of hard work. Value builds wealth. Opportunities sometimes lead to richness. We are being served a huge amount of artificial items: from lifestyle to tabloid icons. How long do they last? Real wealth is for people who build something to last.

We are seven billion on this planet and we can't have an equal way of life for everyone. We've seen socialism failing in action. Material equity – equity in terms of assets owned - can never actually exist, because even if tomorrow at 8am every single person on this planet would have the exact same goods and income, by the time it's 8:30am, we'll already have inequity, as one will start gambling the money, another one will drink it, some other will invest and some other will just enjoy life. That is why intelligent people have a sacred mission to exploit the feebleminded (those that are purely stupid, not mentally challenged in any way), by making them buy things they don't need, with the money they don't have (yet). Modern slavery it's not done through chains and whips, it's done through signatures on bank loans.

Gary Dahl became a millionaire in the 1970s, by selling rock as... pets. Rocks with some eyes drawn on them, ordinary rocks that you could find near a river, sold as... pets. I really hope that these buyers who made him a millionaire, bought stones as a joke, cause this guy surely laughed all his way to the bank. An idea matched in the 21st century only by 'canned air' of famous cities. Next thing you might want to do is find the world's map of stupidity to know where to open up your business.

Pushing to think out of the box, in the same manner you would do this for a product, you can identify the elements that make you a whole. Then consider this: "a chain is no stronger than its weakest link." (which is in fact TOC - the Theory of Constraints). Your most vulnerable link (part of you) can affect, damage or break the whole. If your weak link is your relation with the 'significant other', this is the element that can tear you apart. On the HR side, once this is understood by managers, it might lead to a lot of dismissals. In TOC, people – that includes thinking patterns – are an internal constraint that can limit the output or fulfilling a goal: the way you think, your overall approach is the bottleneck between you and a goal that you have. Jumping back to quality for a second, TOC is highly visible in performance testing where one of the aims is to identify the bottlenecks that reduce the overall speed (response time) of the system.

Traditions and cultural influences of society have a direct impact on the quality of your life. This cultural influence is much easier to fight in places where you don't get lapidated (I wanted to say stoned to death, but the term could have been misleading for younger readers) for opposing traditions. Here is an interesting fact to consider: besides poverty and lack of education, an important cause for children being abandoned or mistreated is the fact that the parents never wanted them. Having a kid was however one of the check boxes on the 'What to do with your life' list, issued by 'Society', placed right after 'Getting married' and 'Leaving this world in a wooden box'. Cultural influences tell you that a residence cannot be a house in the tree or resembling the house of a hobbit and may also decide whether you should or shouldn't buy a car and what brand.

I honestly hope that if you are reading this book, you nodded your head in approval to at least some of the above statements. When I say "I hope", it's not because I would be craving for acceptance, but because I would like to believe I have some empathy towards intelligent people across the world, people who have managed to escape 'consumerism' and the 'assembly line for obedient people'. This was also one chapter not dedicated to managers or marketers.

When did you last manipulate your customer?

Sounds like an evil question to ask, right? Completely immoral. But please note that I'm not asking if you manipulated your customer, I'm just curious when you did that the last time. Well, here are a few tricks that might come in handy, even to prevent from being manipulated yourself. How is this connected to quality? You'll see.

When you are presenting a demo that involves real subjects, assuming you have the possibility of knowing who will be your audience, train it in advance on/for those subjects. If you are presenting a software to a let's say Swedish audience, do expect them to use the special characters in their language when they give it a run. Even if your audience is intended for a specific market – Russia for example, but is said to support all languages as input language and you have one Korean guy checking the product in the user acceptance phase, you can be sure that guy will try native Korean characters. I noticed a sense of patriotism that bumps in during these phases, making people try at least one variation using their home country's currency, address format, special characters, unique identification numbers and so on. Why wouldn't they? Some day they might have to use it, or their friends might and it would be embarrassing for them to say they validated it.

Colors look better than numbers. Software users and people who ever got their hands on any type of equipment learned that, internationally, green means good, yellow is warning and red is bad. Numbers can look worse. Picture this: you are measuring the accuracy of an equipment. Put on the marketer's hat: you establish for the demo, a threshold for being accurate set to higher than or equal to 75% - meaning that if my detection has an accuracy of 75%, I want it to display Status Green (literally a green square or the exact words Status Green). Now put on your customer thinking hat: you are watching the demo and you see that all is green – it's good, right? What if instead of that, you would see the number: 75% - well that means we would be failing by a quarter our accuracy tests and that's not quite something to be proud of. Just be careful not to play with this trick around safety-critical systems, especially in production.

Element sizes in diagrams or presentations can help you draw emphasis on what is of interest for your audience. It can also hide that you plan to do something very complex. Imagine you're drawing a diagram describing connection between logical components of your plan. I'm avoiding on purpose to say this is technical architecture, or software-related, because it can literally apply to anything. So: you're drawing a diagram which, for the sake of simplicity, let's say it has five elements: user – security – processing / manufacturing - auditing – reporting.

From the perspective of a logical structure and logical connection, all these components have the same importance. BUT – here comes the trick – if you are presenting this to an executive, you can double the size – font size or box size - of the element called 'reporting'; for a banking staff, you can emphasize security and auditing; for a technical staff you can enlarge the boxes of 'security' and 'processing'. As this would be a logical diagram, it literally makes no sense which element is bigger, but the target audience will feel more comfortable to see that their area apparently gets the most attention. Try not to make it too obvious though.

Same pattern can be applied in advertising. If you're selling a digital lock, the emphasis would naturally be on security (digital LOCK, ok? not clock – pay attention), but fact is that right now, only early adopters would buy it and these are usually tech enthusiasts. However the biggest fear in the case of a digital lock, is that you could remain locked outside just because it ran out of power, or the digital code fails or the fingerprint reader is not working. So if you're trying to sell a digital lock, the highlighted words would most likely be security, technology and against the "Titanic principle", 'infallible'.

Product placement. We are all aware of it when we see it in the movies, when the coolest guy in the film is wearing some specific brand of running shoes, so why wouldn't we use this technique in our own activity. If you are manufacturing notebooks, client-facing sales personnel should be equipped with top of the line notebooks and trained on how to use their best features. If you're making furniture, the most comfortable chairs and best design furniture should be in the meeting room. If you're a designer, I would expect to see most of the things you are using or wearing, branded by you (from business cards to anything else). I agree that not all solutions are so straight forward. The idea is that product placement is collateral to an activity and not the essence of that activity.

Looking at the examples above: the notebook serves as means for presenting something on it, yet if it's top of the line, it would be impossible not to notice its features. For the sake of amusement, if you want to show processing power you can spread around during the presentation a file that requires heavy resources and ask people to open it – theoretically a file needed for the presentation. Once they fail, you can just say: 'wait, I will open it on my computer, 'cause it's faster'. The comfortable chairs that I mentioned: they are just accessories in a room meant to serve other purposes, but it's impossible to ignore a comfy chair.

Image sells. Suits inspire stiffness, smart casual inspires confidence. At least when it comes to software consultants. Imagine you are visiting the headquarters of a software company and you are visiting the open space or the room where developers, testers and business analysts are. If you see everyone wearing a suit, would you trust it? (Unless it's North Korea) I wouldn't. Developers don't dress like that, neither do testers. What if you would see some sleazy guys, wearing shorts, with gaming t-shirts or coding jokes, sitting in a chair, drinking caffeine, what would you think? Well, it could look believable, but I wouldn't trust my business to some people who can't take care of their one appearance. Then what about smart casual? Well, that sells. It looks comfortable, believable, but still serious, mature and neat. Image sells. If you want to sell consultants, your people have to look like consultants. If you're building only in-house software, then let people enjoy themselves. I've seen companies supposed to sell consultants, where some people were dressing like coming from a barbecue (or going to one). They were definitely working for the wrong company (no matter from which side you look at it).

The image sells, but the inner quality retains the customers. People don't always remember who helped them, but they always remembered who disappointed them. You might be able to fool them twice – shame on you, your reputation is gone.

As a consultant, **you help to improve, never to correct**. Consultants are there for improvements, not to solve problems, cause there aren't any. I've worked on a few projects where old products were being refactored. Fact is the old products were junk, but you wouldn't hear the customer say: 'We messed up so bad, we need a full replacement'. The official version is: 'we want to improve our product by creating a new, more modern version and we will sunset the old product' (I love this expression with 'sunsetting software' – it gives it a romantic – dramatic twist).

By the way, if you remember what I said a few chapters earlier about the enemies of quality, this is one example of faking. No one likes to see other people spitting on their work, so whenever you feel like saying bad words like 'inefficient', 'poor', 'barely usable', or synonyms, dip your tongue in honey, think a bit, then try again. Here are some suggestions: a product is not poor, it has a limited number of functions; a product is not barely usable, it addresses more experienced users and it needs adapting for mass consumption. The technology is not old because some cheap bastards used old technology or unskilled personnel: it's old because technology evolves at a dazzling speed and the company wants a refactor, to keep up the pace.

It's cheap, fast and very good!...said **no customer, ever**. But sales people say it quite often. Whenever you hear the above phrase you have to figure out which one is the lie. However, expensive and slow will not lead to quality - it could just mean you have a serious problem in production. This particular unlikely conjunction of features is what makes me avoid buying products freshly released around Christmas or Easter. This excludes by definition products marketed a year ahead, to be released around these dates. It's natural for a company to target these major events when people open up their wallet with generosity, making a good deed by buying merchandise from a major electronics producer, but I also know how things work on projects.

That release prior to major Christian holidays (could be any other religion or event for what it matters), is focused around key dates and milestones. I also know that due to optimization, tasks were brought in parallel into the plan so the whole marketing campaign started before the product was actually finalized. Since the manufacturing process takes time, this also has to start at a given date – another milestone. Prior to all these milestones, there is one called 'test exit' and another one called 'development freeze'. Between these last two, there is a pile of defects in need of fixing. As we like to joke in testing: 'every bug's dream is to get into production'.

Unfortunately for them, fortunately for companies, **people often buy the label, not the merchandise**. Rephrasing: mass consumers buy the label, not the merchandise. This can explain why counterfeit whisky is quite successful: few people know what it should be like and rely on the label instead. Problem is they can't tell the difference, not even after tasting it. This consumer behavior also explains the temporary success of branding perfumes, headphones, all sort of stuff with the name of music stars and no other particular brand behind them. I could understand and believe that a pair of headphones branded by Dj X and powered by JBL or Panasonic is a good brand, since in fact this is just advertising. But it's hard for me to believe that Dj X powered by no one else but his own investment, learned over night how to produce (manufacture) hi-fi stereo headphones or any other piece of material involving circuits and technology.

As a homework, if you want to experiment consumer psychology among friends, you can take some ordinary wine and stick a fake label saying something like 'rare selection, limited edition, black label, aged in 'some-special-wood' barrels and say you got this as a gift from your office - so that people will be tempted to examine it. You might notice that this wine will be received as *'much better than the ordinary one'* which is in fact what you are actually serving.

I did a little experiment once, praising a house music band from Denmark (seemed like a good choice to say Denmark, as not many people would be familiar with their music). This was a band which I made up, exhibiting some music sample created by a friend of mine on the computer. I might've said playing the sample to some friends, that they got some best new comer award and that it was the best underground band right before that. Can you guess what the reaction was? Some people said not only they knew the band, but they've also seen the video. Well, these people didn't get to listen to the album, but they were removed from my circle of acquaintances.

People buy what they are told to buy. This has nothing to do with reading the labels or the specs. Advertisements matters a lot because they make things familiar. Psychologically, people feel more comfortable buying things they are familiar with, even if they actually don't know anything about the producer. Then, of course, you have the people that don't read the labels anyway, so whatever they've seen on TV must be good if the producers afford to constantly advertise. I believe this is visible when it comes to comparing mineral water brands.

Let's assume we are not talking about anything with special characteristics – where the consumer might be interested in a particular trait. Let's say one of the products is "Sparkling DeLuxe Water" (I made up this name, I have no idea if this brand actually exists) and is highly advertised, while the other one is "Anonymous Sparkling" (the name speaks for itself).

Even if the content in minerals is identical, most buyers would pick the first one, because they simply 'know it' even if they don't know the history of the brand or how good its content is. Besides the feeling of familiarity, there's a lot of manipulation also. Soap ads and over-the-counter medicine ads (for which you don't need a prescription), they scare the users with lines like: "there are a lot of germs around, you need to use this soap to stay safe, since this is the only soap that obliterates 94,6% of the germs" (see how the numbers give you a feeling of accuracy and expertise), "as you get old, you will have ..... health problems" (you can add any word on the dotted space), "you need to take this medicine early to ease your problems, before it's too late".

There are other forms of manipulation that expand beyond customers and support production instead. To bring in an atypical exhibit, I would have to ask you: is caffeine an investment or a cost? Let's face it, caffeine is a drug. And I'm a complete addict to it. That doesn't make it less of a drug – it enhances some senses, enhances perception, the organism of an addict feels bad without it and needs its daily dose. If it were my decision, I would make an effort to give free coffee to employees or at least subsidize the price on vending machines around company's perimeter. Modern slaves – the obedient office workers - need caffeine. It stimulates their attention and thinking and there's a smaller chance of making mistakes. Think of it as upgrading your workers – I would want such an upgrade at a low cost. Enhanced attention to details raises the overall quality, plus it's a benefit given to employees, which in some countries might be tax exempt.

It's difficult to place employee benefits under the umbrella of this chapter, because in the end it's a fair trade for retaining a good asset. You give the employee a good phone, a company car, some stock options, some discounts for the duration of the contract, so why would someone leave knowing he/she has to give back all the candy? Well, it works until someone comes with a bigger candy.

The farming and leather industry, however, learned how to trick customers by **changing the wrapping and make it sound beautiful**. Many so-called 'modern' farms rely on intensive exploitation of pigs, cows or other animals, raised on concrete floors, which have never seen grass under their feet, growing in boxes where they can barely turn, that if their head isn't trapped through metal bars. A sane person cannot enjoy seeing that versus the traditional way of raising cattle. Yet this is presented as an "intensive effort to improve productivity" (sentence which makes you think "progress" – cause they are improving productivity, right?). In the ads or on the labels you can see happy animals, smiling, as if the end product that you might be buying didn't end their life.

Similar to this, the leather industry tells stories about hunting (as if there was some sort of bravery at aiming a defenseless creature from a mile away) and advertise (!) with live animals, when their product implies skinning them. So these manipulators teach us a thing about human nature: with the right label, wrapping or pretext, you can mask even a barbaric act, for the simple reason that the nature of most people is cowardice and they prefer to turn their head away from (or hide from sight) things that they know they are not right.

Paper industry does not show you barren hills with trees cut from the root, they show you kids studying in school (they support "education" – it's good, see?), they show you teenage girls writing in diaries, giggling or the ambient of peaceful home (see how peaceful things can be if you have agendas and notebooks?) or business men taking notes at a meeting (you must get this product if you want to have a career!). Honestly I can barely remember the last time I wrote a full page on paper – it must've been during Uni – since I nearly always had a computer around or a smartphone to keep short notes or obtain information without having to write it myself. For the same reason, when I am asked to fill in large quantities of forms at a bank I may seem a bit illiterate, since in my day to day activities, I rarely write more than three words with a ball pen.

A good quality product with a sane production chain, despite using all artifacts offered in sales, including different forms of manipulation, to increase profits, **has the power or possibility to expose the full chain of production, making it transparent**. This has nothing to do with intellectual property or patents, or manufacturing secrets: it's not about exposing secret recipes on a wall. I don't think a manual watch maker would have a problem showing you a guy sitting at a table and creating a masterpiece. A gold manufacturer instead could never show you how a miner with minimum wage went down in a mine and worked 12h a day or how they used cyanides, poisoning lakes, in order to be able to create that beautiful ring for your beloved one.

A typical and more excusable way of manipulating buyers when it comes to technology is based on **gender psychology**. White or pink cases with flower models on them (or similar ornaments), basically looking beautiful will draw more feminine customers as long as the quality is reasonable, or they would even pay extra for this, because a beautiful product makes them look more beautiful. Meanwhile male consumers will care less about pure esthetics (which is a nice-to-have, if possible), but more about the technology inside, giving them the chance to say 'mine is faster' / 'mine is stronger'. This sort of approach works for products that are individually owned (like a camera, laptop, tablet, headphones, mouse) and preferably portable. This doesn't work well for TV-s because that one belongs to the 'household', to the 'family' – you can't associate it with something personal.

So here we are, at the end of a chapter from the darker or more intelligent side of sales (a matter of perception). Let's see how the world is seen through the eyes of a tester.

Tester's amusement park

Instead of further preaching about the importance of (not) doing things right, I'm going to tell you how the world is seen through the eyes of a tester. Because once you start a career in this area, you become attached to details, noticing logic flaws, asking more 'what if'-s and 'how do you know'-s and if you have the slightest interest for security or penetration testing you're tempted to push the limits and cheat the system.

All the stories you will read relate to my personal experience or to that of close friends.

Testers know: A bug is a social insect. You'll never find just one of them. Wherever you can spot a bug, there are many others to be found or at least its mate, if you take the time to look for it.

The never-idle-support-officer

Before starting my career in an official quality control role, I used to work in technical support. I started off by working for an antivirus producer, responding to all sort of requests by email or chat. That was a place where we were actually struggling for quality – it was a young company wanting to conquer the world, all employees were young and willing to prove worthy. There was a lot to learn about computer viruses, services, workarounds, registry entries. It was there where I became interested in quality. So following this path of testing, I moved on to another job where for a few months I had to do some more technical support, this time over the phone, until the position I was targeting (an entry position in testing) became available.

Honestly, this support over the phone was driving me crazy. In a world where Adam Smith published his theory of the Division of Labor showing the advantages of having specialized personnel (and we're talking here about a theory published in 1776) and Tocqueville insisted on the idea of decentralization around 1820-1850, more than two hundred years later, some companies come with brilliant ideas like SPOC support. This means a **S**ingle **P**oint **O**f **C**ontact, where a bunch of Mr and Mrs Know-It-All will answer whatever question you may have.

In my case, this meant that any of the persons working at this call center (I think it's a more appropriate name rather than support center), was supposed to know everything about a few hundred products ranging from printers to cameras, scanners, etc. After each call you had about 30 seconds to finish typing your ticket, until the next call would come in. If you felt the need for a break (which couldn't be longer than five minutes under any circumstance), you had to select on the phone the reason for leaving your desk, explicitly selecting if you went to toilet or to lunch, or another predefined answer.

As if this wasn't enough, about a month after I started the hands-on activity, management decided to mount a projector that would show real time the status of each person, the response time, SLA targets, etc. This projector was part of the system controlling the calls showing also who was the person spending most time idle and who should take the next call (although this was all automatic).

Watching this screen for hours, while doing my job, I noticed there was something weird with the refresh time of the 'longest time idle' area. So for two days I monitored what was happening when I was going on a break and that's when I noticed the aspect that restored my peace of mind, for the next month, until I moved to a full time testing position. Here's what was happening: when pressing the resume button (coming back from a break), the system was resetting the idle time. Therefore, immediately as you returned, you always had 0:00 seconds since you last took a call, so you were becoming the last person who would pick up a call. Of course this was a logic flaw, because since you were returning from a break, you were actually the fit candidate to take the next call.

Due to a glitch in the state transition diagram, the system was not distinguishing between a person returning from a break and someone who just finished a call. So I thought: ok, this would be a nice way to cut down on my work, but putting myself on a break (in order to be able to 'resume' and reset the time) would've shown me 'away' on the big screen. Since supervisors were watching this screen like it was their favorite sitcom, I didn't want to appear 'away' on the wall.

During my two days of observations, while working, I also noticed that the system had a visible lag in another area (visible for a trained eye at least). Firing a few tests on the desk phone, I confirmed my theory that the system was not picking up the change in status immediately. In fact it needed more than one or two seconds, which was a generous time frame to press fast enough the 'change state' and 'resume' buttons. When I realized all this, by confirming all my theories on a world-renowned brand of desk phones and networking equipment, this became my favorite game: pressing as fast as possible 'change state' and 'resume'. My status never changed on the screen, but my time was always reset. And from that moment on I started receiving half the calls my colleagues had, turning a situation I was finding miserable into a bearable situation for another four weeks.

If it mustn't be on the bill, but it is there, it means you don't control your purchases

For a tester, seeing a 0.0001 discount printed on a bill means you had lazy developers or cheap ones. Either they had little time before deployment to do all urgent fixes or they simply didn't know how to do it. I once saw a bill, where after the subtotal, you had a -0.0001 discount, which was obviously pointless. Why would it show up, particularly in this form, when a simple rounding, could've made it show just -0.00

This tells me as a tester that your system did not implement properly the situation where you simply don't grant a discount or discard it, if by rounding, it equals zero. It also shows me that the developers did not want to spend time cleaning and correcting the printed bill. So, if I were to start an exploratory session on that app, I already knew where I could find some defects and prove myself worthy: discount scenarios and printing functionality. If I exploited this further, I could assume that your system supports discounts (as most modern billing apps should) and there's a high chance to be able to trick your purchasing app into granting a discount, by intercepting server messages or by editing some possibly hidden fields. One single drop of blood is all a shark needs to find a victim.

The high score secret

Game high scores can tell you if an application (a game) has serious bugs and also how many people found out. When it comes to games, we admit that some people play games as if it were a full time job, and do unpaid overtime. A sports publication I was reading years ago – when they actually had some good titles once a week – was running an online game with some pretty nice prizes.

The game was meant to simulate the internal football championship (note to American readers: yes, it's called football, not soccer). You were acting as the manager of the team, had an initial budget, played against team of your own caliber and could increase your budget to buy more expensive players. The winner was the one having the highest amount of money, at the end of the official championship. The players you could buy were rated (and priced) based on their real performance in the championship. You could only play five games a day against other teams, for free. If you were buying some designated products, you had an extra game that day, based on a promo code.

Only a few days after this game was launched, I was seeing some spectacular scores. Doing some easy math, I reached the conclusion that if you won all games, and buy all products granting you extra games, and win those as well, not even then could you reach those scores. So that is when I caught the scent of a bug. A security tester (penetration tester or white hat hacker) is like a hound. If your product is limping and it caught the scent, he will chase it and dig for it until it finds it. I wasn't even interested in the game anymore; I wanted to find that flaw.

First hint was that the only way to make money in this game was by winning the daily games against opponents. These games allowed no manual intervention during the play so it was something between lottery and algorithm. Next thing, I started recording the http session to see what happens – maybe the application was silly enough to send the budget numbers without any server validation. It wasn't the case. Next thing, I thought of validating whether the server is registering each game as a unique session and this... was not the case. When I replayed the recorded session, my earnings doubled. Tried it again, it tripled. I went in ten minutes from a budget of a few thousand credits to a budget of millions.

Of course, the game had all sort of legal disclaimers, forbidding cheating in any way, but nobody seemed to care too much – not even the organizers. Well, there's a thing I don't like and that is playing unfair games, so the moment I found this I played for a few more days, interested more in exploiting the flaws, but then I lost all interest in the game. There's no fun in playing with a broken toy.

Shop online if you can

While doing some online shopping, I came across two websites – amusing in their silly nature, but not the places I wanted to do business with. On the first one you were required to create an account in order to buy, but the account creation process was damaged and you couldn't make an account. Funny thing is: if you would tell the site owner that he has this problem, he would take an existing account and try to do an order. What's the connection? Beats me, but that's how some people react. What about new users? Anyway, so this one was quite silly and most likely they went bankrupt since they didn't receive any orders.

The second website I want to tell you about was not sending the buyer a confirmation email. At least, as a buyer, that's what you could know for sure. But it wasn't sending an email to the seller either. So, in fact nobody knew you placed an order. I waited for two days to be called for confirmation and since Christmas was approaching, I called them. To my surprise, I was nearly scolded for not calling them after placing the order, in order for ME to confirm that I placed an order. Obviously these guys had nothing in common with web commerce. Surprises are everywhere – I found also websites where you couldn't add a product to cart, yet they wanted to do e-commerce. As you can see, even in the software environment, there is a natural selection process.

Cheaper racing offline

Less than six months ago (prior to publishing this book) I discovered a car racing game on the mobile phone. I usually try to avoid installing games, because I end up wasting a lot of time by my standards, but sometimes you really need to have something to fill some dead hours or minutes, particularly while waiting for something. It would be better to read a book, but I don't always feel the need to do something useful. Anyway, what attracted me in the first place were the HD graphics and some good looking screenshots. The game was quite responsive to phone movement, so I was really enjoying it.

Since even games are business, you have the chance to purchase all sort of in-app upgrades, car packs, boosts, to help you throughout the game. What I appreciated about the game is that it was also giving you a fair chance to advance in the game without buying anything, as long as you would play long enough to earn credits. With the credits you can buy other cars, upgrades etc. Game can run both offline and online. As I don't always want to receive push notifications on my phone (like emails, app messages), I turn off any data connection when I'm home. This game can record your status on the server, only if you are online, obviously. So at the end of a play session I turned on the data connection, just to upload my progress, but I didn't disconnect it immediately after.

While doing this I noticed two things: the cars available for purchase were changing when going online, and so was their price. So, as any tester would do, I tried to test and isolate the defect. Indeed: if you were purchasing offline, without going online at all, after entering the game (this was very important), all upgrades and prices were 2-5 times smaller. Also for some particular cars, price was dropping when going online. This difference in price could have been as much as 200 000 credits, which would've taken more than a week or two to obtain. You can imagine how my shopping with game credits went since I discovered this.

This was also telling me that the game is not validating against the server any credits or progress and since progress is kept from one session to another even if you don't go online, it means it persists, it saves this information locally in a file. So next thing I tried was to hack the files to see where this is stored. I wasn't that successful in this aspect, mostly because I gave up after half an hour, since it wasn't of that much interest to me. I already knew it's in there 100%, it just wasn't worth the time. The game had two or three major updates (1Gb updates with full reinstall) since I found this problem and still no fix for this bug. Out of professional courtesy, I intend to email them as soon as I finish the game – but I'm not in a rush to do so.

Solitaire on ATM

As much as I would like to share this story with you in full detail, it could pose a threat to persons I know. So therefore I will just say, that if one day you go to your bank to withdraw money from the ATM and instead of the regular welcome message, you'll find a 'Solitaire' game displayed on the screen, it's because someone was too lazy to do a good coding job and because the software is running on a Windows platform (Windows Embedded POS Ready) which sometimes does not fully extend when going full screen. In this kind of situation the finger nails of a lady might accidentally tap the lower left corner of the screen and bring up that Start button... Well, enough with this.

Sensors

There's nothing you could trust less in this world, after governments and television. A sensor can be tricked into reading anything. Problem with sensors is that they are not only reading data, but they try to make decisions based on the readings. A friend's car was set not to start the engine if it didn't detect gas in the tank. Therefore, when the sensor broke down, although there was enough gas in the tank, the car had to be swung a bit to detect the gas. So before taking the car to a garage, the method to start this car was to ask a friend to bounce on one side of the car, while he would bounce on the other, so that the sensor would pick up the gas and start the ignition.

General poor quality of sensors is the reason why my (theoretically life-saving) gas leak detection sensor was ringing every time it would feel perfume or even the smell of cooked food.

Motion laser sensors are fun to play around with mirrors. Sounds like a James Bond scene but it actually works. I was working once with robotic equipment which by industrial standards needed to have some detection sensors for emergency stop if someone was to close while it was operating. When you're working in a lab and want to do constant measurements, moving around, the last thing you need are these 'dumb-ass protection' features. These sensors consisted of two bars (about one meter length each) with about a dozen emitters and receptors on each side.

These bars had to be perfectly aligned in parallel, so that the beam from one bar would reach the receptor on the other one and send back the beam. As soon as the guys from the manufacturer installed the equipment by all standards, we brought the bars down and put them next to one another and started playing with mirrors. As a side-note, it's a common prank to stick reflexive material (like transparent tape) on the sensor of an optical mouse.

Motion-triggered detectors are generally pretty bad at detecting very slow movement. Some hotel card readers placed inside the room for activating electricity don't ready any information at all, they just need a piece of plastic in the slot, which can be your gym membership card just as well.

Given these examples, whenever I encounter something with sensors, I am more interested in discovering where they would fail – it's better to know in advance.

Pizza delivery bug

One evening, eager to order a pizza after I found a leaflet at my door, I noticed a reference to their online app. Curious by nature, being my first pizza order via Android app, I tried it. I set my details, selected the order, composed a custom pizza, all went well and easy. Easy as...pizza, until I received the confirmation message, which came as a push notification, showing a message on the screen displaying exactly the following content 'Your order was processed and will arrive in 3...' (dots at the end are part of the original message). That is how the visible part of the message ended on the phone. I pressed on the notification and it opened the app where the message didn't pop up and there was no section for received messages.

Of course I didn't assume pizza is going to be at my place in 3 minutes, hopefully not 3 days either and I definitely prayed they didn't mean 3 hours. So I assumed it was saying 30 minutes (which was a good prediction), but an embarrassing defect. Funnier was that when sharing this defect with two other fellow testers, telling them about my first experience with the app, all in good faith, using it by the book, each of them told me about another defect they found when trying to edit their details or send a message.

Circular logic

This example I discovered in one of the office where I worked, when trying to configure my wi-fi connection. I have to say that due to security restrictions in place, activating the wireless connection was not as easy as 1,2,3, like you might be used to on your personal computer. To use the company's wireless connection, you had to configure a security protocol through dedicated software, but the guys that worked on the tool decided to place the help file... on-line.

Generally speaking, how many people do you know that try to configure a wireless connection, while they are already connected by cable (and therefore able to access an online page)? I surely wasn't among them and this is a clear example of circular logic: go online to configure your wireless connection, through which you would go online.

Take the money first, ask questions later

This one refers to a coffee vending machine, which I'm sure it brought consistent income for a short period of time, since it was simply a money-eater. This metal box did not have any sensors to signal if the product was delivered, if there are enough ingredients to deliver the product or if it should return any change. To exemplify: if you wanted to buy a 20 cents coffee (subsidized price, never mind) and you've put in a two euro coin – you just got yourself a two euro coffee or ten normal coffee cups.

I had to find this out the hard way, with the machine swallowing the coins and then showing 'unavailable' for every single product selected (yes, I tried all of them, wanting to get at least a cup of tea from that greedy metal bastard). That's why people start hitting vending machines. A common failure on the machines that have delivery sensors is that they place the sensor too high above the delivery tray, and products can still get stuck.

IOT - The Internet of Trolls

Only two months before finishing this book, I received an email from my cable tv provider, in which they were proudly announcing the release of their mobile app through which you could record shows and change channels from your phone, if you had their DVR HD system. I thought – well, this is a nice way of exemplifying the Internet of Things. So I got the app, installed it on my phone, which at the moment was one of the nicest things made on Earth – a pretty good device, ready to handle mostly anything that was on the market. I accessed my cable tv account via phone app – all good, it even recognized my equipment at home – it was all looking beautiful, until I tried to use the features advertised which were the essence of this whole effort: recording a show or changing a channel from my phone. None of these worked.

As a nice tester that I am I thought of emailing them, hoping they would do a common sense gesture and release a hot fix in the next two days. Not only that never happened, but they made an appeal to my understanding saying that the app is still "under construction" and has some problems, when the whole product was in the app stores for over six months, according to timestamp in stores and advertised in their last newsletter.

Apparently this company asked its employees or maybe alpha testers (users doing acceptance tests in a controlled environment) to give it a 5 star ratings in the app store. So there were nearly 300 rating saying mostly nothing – comments like 'excellent, best app ever' or 'simply loved it' – and 120+ one-star ratings complaining about specific issues like the ones above and calling the app 'garbage'. Regarding their manipulation attempt – *thanks, but not today.*

So you see, a good tester finds this sort of faults everywhere. Most testers I know have found algorithm failures in modern elevators, on the selection command – this is not because we tested it, we know because we noticed it and at some point tried to validate it.

For us vending machines are devices that detect metal weight and size and that can take care of you if you give them the correct volume and density of metal. We are used to spot deviances from patterns and we have analytical thinking. We can't watch a movie by ourselves, without replaying a scene if we think we spotted something that didn't match (even if it's a director's cut). As for security, well... security starts from design. Any visible weakness is like a trace of breadcrumbs leading to an open door. Even on the most secured systems, electrostatic discharges on physical ports can have funny effects on machines.

Everything around us can be part of a huge playground if it gets our attention.

The rotten apples

Under the umbrella of this title, I decided to gather at the end a number of bad practices and bad examples that in many companies have become business-as-usual. You can't have an apple tree grown without pesticides, fertilizer or simply without caring for it and not get some rotten apples. 'Laissez-faire' policy comes with the cost of inaction mentioned in earlier chapters.

So here's what the basket with rotten apples has to offer:

Accepting a (software) crash, because 'few people will get to it'

No. A crash is never part of a software. This statement should be enough. Even if employees with lack of self-esteem may accept it, a crash visible to the user under normal usage conditions, no matter how rare would be the concurrence of factors that led to it, transmits one and only one message: poor quality of coding. Let's say you see a crash in your internet banking application - would you still trust your money to it? What if you would know that the tower control equipment guiding the airplane you are about to board occasionally crashes? Under really rare conditions of course, but nobody fixed it, because it seldom happens. Would you still board a plane leaving from or reaching that airport?

Acceptable failures

Failures are never "acceptable", unless you label your product as mediocre. Testing everything is impossible – it's one of the principles of testing. Also perfection is hard to reach. So we accept that during the manufacturing process we might see some faults (failures). The only one that can decide whether they are acceptable or not is the user. Example: For many years, it was a common policy among LCD display manufacturers (this covers from phone screens to displays and TVs) that one dead pixel is a tolerable fault and is not a reason for product replacement.

This has changed in time, for some producers and as screen resolution has evolved, noticing the 'one dead pixel' became harder. (For those of you not knowing this concept, a dead pixel is - simply put – one dot on the screen that either doesn't lit or remains continuously lit no matter what you display on the screen, including a black image).

So it happened that I once had a phone, before the era of smart phones, when pixels on the phone screen were a small group where everyone mattered. The moment I got a dead pixel, especially one close to the center of the screen, I can assure you that was no tolerable fault in my perception. In fact it bothered me so much that I never wanted to see a phone from that producer until they change their mind about acceptable failures. Come to think of it, after this event, I never bought again any other product from that manufacturer although I didn't make a commitment to this, but it's hard to say if my subconscious decided that it was 'bad' to buy things from that producer, based on my passed experience.

Giving up on defects because you're the one and only tester

This happens when the rest of the team is putting pressure on the only team member acting as the user's advocate. If you as a tester know what you're doing, it shouldn't matter and you should stand your ground. You might be called stubborn or arrogant. Justified arrogance is called confidence, if you can actually back it up.

When you finish the tests, you have to give a feedback. That feedback is not always positive. You can at least earn your peace of mind by knowing you honestly reported every problem encountered. The final decision – the emperor's thumb up or down – is not yours and the product is not a one man show.

You do need a strong personality to handle this and a big pair of... arguments. Whoever has a problem with the defects reported will have to prove they are not real. Everybody involved in the development of a product knows how this game is played and you should know that you're the only one who will be accountable for the test exit report.

Being the opossum tester

I would like to give full credit for this expression to James Bach. I don't know if he invented it or not, but I heard this term for the first time at one of his conferences. The 'opossum tester' is a quality control person who executes some work that he/she knows it's useless. This would be the definition in a nutshell. In fact if you feel that this phrase describes what you do, you shouldn't be there at all. Quality control personnel need to have personality and argue over their findings, in the sense of defending one's point of view and not just being stubborn. Obedient staff in this area that simply says 'yes sir' to whatever the project manager says or accepts blindly directions given by others, this sort of staff is just a waste of budget.

I've seen projects (fortunately I didn't work on them) where testing was in place just for the sake of having it, just to justify some costs, and whatever happens after releasing the product, well it will be just bad luck. If I were the client and wanted to know how my product is doing, I would like to speak to someone from QA. Since it's highly unlikely to have the privilege or the occasion of a one-to-one meeting with a person working on execution, no one can refuse you if you want to meet the team building your product. In this situation, as a client, I would ask the QA person only one tricky question: "If you could change one single thing in my product, what would that be?"

Not taking an extra step can keep you safe in mediocrity

I once worked for a project where a login based on OCI standard was used. Imagine the whole thing as a link in the address bar of a browser with several parameters. In the beginning only four parameters were used – easy to handle, but it was annoying to build the login from scratch depending on the test. Since I had some extra time on my hand, I made a small app where you could edit the whole thing in a UI, press a button to load it to clipboard, and all sorts of other small features.

All this was done as a hobby, mostly for myself, just to simplify my work – because as my father-in-law says: "*if you want to find the easiest way to do a job, give it to a lazy person*". So from the coding perspective, since I had just four parameters, I wrote some IF clauses (some ordinary conditions saying 'if this happens do this, else do that') and copy-pasted them, cause it was the fastest way to build it.

This was not the best or most correct way to do it – but hey, I was the end user and I didn't care that much as long as it did the job. Some weeks later, those four parameters turned into sixteen and about two months later, or faster, they reached forty supported parameters. What better way to add them fast, really fast than keep on copy-pasting.

About five-six years after doing this little app, I was interviewing a candidate who, to my surprise, was working on the very same project that I worked on, five-six years before. So asking all sort of questions, the candidate mentioned OCI login standard, so I couldn't help but asking: *"And how do you compose this login? Do you type it manually every time?"*. This is the moment when the guy said: *"No, we have this little app made by someone who worked on the project, in the beginning, and we use that"*. I honestly think it was the lousiest piece of coding ever done, but still in use, just because no one else bothered to do something better or anything else at all. Sometimes all you need to do to make a difference, is to take an extra step.

"Chaos model"

Incompetence has reached such levels that the notion of 'chaos model' emerged in software development, developed rules (!) and even started to act as a legit model. Originating in the mathematical chaos strategy, this so called model launched by someone who avoided releasing his real name has in essence one simple rule *'always resolve the most important problem first'*. Its strategy considered defects as incomplete tasks while the importance of a task is given by size + urgency + robustness, where something is urgent only if impacting other work. The resolution in this concept did not mean complete fixing, but just bringing it to an acceptable level. A lesser evil is the 'code and fix' model which in essence reflects the effects of schedule pressure, the name being self-explanatory. This latter model is known in day to day life common terms as 'patching'.

Self-training

Two-day training can be worth more than a week of self-training. Specs are seldom reliable for learning. Self-studying on how to build nuclear reactors, may give some results...in a decade or two. Any learning phase (sometimes called transition) can be significantly shortened or made more efficient by bringing in a trainer. Yes, it could cost extra, but a trainer is not normally hired for one-on-one sessions. So you can have five people self-studying for two weeks and you end up paying for 400 hours, and each one will understand something different OR you cut let's say two days of self-study, meaning 80 hours and pay instead for 8 hours of training, getting the same content delivered to everyone, by a professional AND explained. It's pure budgeting and fourth grade mathematics.

Not giving access to testers on a productive system or to an end product

This happens more often in software development and seldom when it comes to assembly lines. Let's discuss first the latter: in some industries quality control on random samples is a must and if we're discussing electronics, the testers might be quite passionate about what they are doing and be eager to get their hands on the end product (or simply buy it, if the price is reasonable). This obviously can't happen with very expensive equipment. I worked on a project many years ago, where we were building software that had to integrate with a high number of household electricity meters.

Despite our repeated demands, just to be able to build our own testing panel, it took more than a year before we got our hands on few of those meters. A more common situation when it comes to software is that of preventing the tester or at least the test lead from having access to the production system, even if there are confidentiality and non-disclosure agreements in place. This stops the testing from tracking some errors that manifest only in that environment, cuts off the possibility to investigate production problems and causes a hot fix delivered in production to rely up to 90% on tests done in other environments and 10% on faith.

Constant refusal to learn from past mistakes

According to different philosophies, whoever does this is condemned to repeat them forever until the lesson is learned or the game is over. A friend of mine had a commercial space to rent in a suburb of a big city (we could call it a small town). Three times in a row, the tenants (three completely different persons) decided to open a bar with a small terrace. All of them went bankrupt. When the fourth tenant came to rent the place, my friend asked out of curiosity and perhaps for legal reasons as well:

"So... what do you plan to do here?"

"Open a bar, of course"

In good faith, my friend felt compelled to tell him the story of the other three tenants that had the same spark and failed miserably. The answer was: *"No, this one will be different, it will work"*. Not only it didn't bring anything new or different, it shut down faster than the previous three, in about three months from opening.

Nowadays, I am noticing a trend in the mobile gadget world to bring out intelligent watches. Not only some of them are as expensive as a smartphone, but the producers refuse to acknowledge basic things: a watch is small by definition. Even if it is a bit bigger, without making it look ridiculous, our fingers aren't shrinking and might be quite fat sometimes. Therefore to maintain usability, the elements on screen need to have a certain size, which could reduce the usable surface to four-six elements.

This cuts down on the number of features since having to go through four screens to perform an action that could be done with a single tap on the phone screen is simply annoying. But maybe the most important, tapping on your watch or even speaking with it, doesn't make you look like a secret agent or a SF character, it makes you look stupid.

Pulsar made the first smart watch in 1972 and they had far less things to display. Casio made 'computer watches' and 'calculator watches' in the 1980s. In 2000, a model running Linux, with Bluetooth, accelerometer, fingerprint recognition, vibrations, was made and discontinued in less than two years. Sony's smart watch is considered by many electronic review sites as a failure. A friend of mine used to have one of the Sony watches and I am not saying that the technical specs were not good – I'm quite sure they were ok, but he definitely looked stupid when trying to use it: struggling to type on the screen to see what notification he had, then realizing he'd better use the phone and in the end switch to Bluetooth headset to make the actual call.

More than 20 years ago I had two smart watches, the smartest they could be in those days: one had a calculator, the other one had a radio incorporated and a connector for headphones. A school mate of mine had one with games on it. They all had the wow-factor for those days, and they were all more annoying than anything, to use.

If the element of novelty was there already and it didn't explode, accept the facts.

Pushing for feedback

Collecting online feedback about user experience immediately after accessing a site and making a twelve-page interrogation is just a method to scare people a way. It feels as if you've just met a person, and the next day that person wants to move in with you – if you're not comfortable with this idea, you can understand why people are not comfortable with feedback pop-ups immediately after accessing a site. In the end you know who answers these forms? People who are really bored, overly-attached or that feel an urge to express themselves. In a century where a desktop Internet page, not loading in one second, is closed by users, they definitely don't have the time to chat on your form.

Automated translations

Never auto translate your content, whether we're talking technical specifications, help manuals or labels. Auto translations of words and phrases taken out of the context are really poor and the users will know what you've done. Paragraphs auto translated sound sloppy. Particularly, localization labels (like buttons 'names or two-word stickers adapted to a local language) are terrible. Computer translators have difficulties using the plural and the proper articles, but most of all can't pick the right meaning of a polysemantic word.

There are so many examples, they could fill an orchard.

What matters in the end is to care more about the customer, about making something good and once you start creating real professional value, material rewards will follow naturally. True story.

PS:

Dear reader,

Do you remember the chapter about manipulation? Do you know why you read this book? Maybe you are a manager looking for inspiring ideas; maybe you are a marketer wanting to learn more. But you definitely think yourself as an intelligent person and felt like this was addressed to you.

Don't worry, since you have an interest in quality, you are more intelligent than the average and I hope you enjoyed reading it. If you felt like going back through some ideas, I will take that as a compliment. The best thing you can do is remember some of the ideas here and apply them even in your everyday life.

Good luck!